ROBERT MITE
PROJECT ENGINEER
MITEC WIRELESS INC
31 PARK RD
EATONTOWN NJ 07724-9716

Feedforward Linear Power Amplifiers

For a complete listing of the *Artech House Microwave Library*, turn to the back of this book.

Feedforward Linear Power Amplifiers

Nick Pothecary

Artech House
Boston • London

Library of Congress Cataloging-in-Publication Data
Pothecary, Nick
   Feedforward linear power amplifiers / Nick Pothecary.
      p. cm. — (Artech House microwave library.)
   Includes bibliographical references and index.
   ISBN 1-58053-022-2 (alk. paper)
    1. Power amplifiers. 2. Amplifiers, Radio Frequency.
3. Feedforward control systems. I. Title. II. Series.
TK7871.58.P6P68 1999   99-26568
621.381'325--dc21   CIP

British Library Cataloguing in Publication Data
Pothecary, Nick
   Feedforward linear power amplifiers. — (Artech House
microwave library)
1.Power amplifiers 2. Amplifiers, Radio frequency
3. Feedforward control systems
1. Title
621.3'81535

ISBN 0-1-58053-022-2

Cover design by Lynda Fishbourne

International Standard Book Number: 1-58053-022-2
Cataloging-In-Publication: 99-26568

10 9 8 7 6 5 4 3 2 1

To my family and friends

. . . the only thing we require to be good philosophers
is the faculty of wonder . . .

Contents

Foreword

Feedforward Linear Power Amplifiers

Communications has made great strides in the last 50 years, in quality of course but even more spectacularly, in the *quantity* of information exchanged.

From the very few TV programs transmitted over the air to the hundreds of programs available via cable or satellite. From the simple information exchanged on "walkie talkies" for almost exclusive professional usage, to modern cellular telephones for everybody. From simple data being printed on noisy and slow teletype writers, to faxes and to electronic mail. From the very confidential consultation of specialized information banks, to the almost real time access to any information anywhere through the Internet. The name of the game is now capacity and speed.

Today digital processing makes it possible to strip down almost all redundancy from the information transmitted and thereby increase capacity. However, we know from Shannon[1] that the ultimate capacity of a communication channel is limited by frequency, bandwidth, and signal-to-noise ratio. The physical separaration of communication channels that was inherent to wire transmission is more difficult in radio communications and makes frequency reuse a difficult challenge. Radio communication systems are therefore confronted with the scarcity of the frequency bandwidth resource. In modern systems, more complex modulations are being used to increase the

---

1. Shannon, C.E., A mathematical theory of communications, *BSTJ*, Vol. 27, 1948, pp. 379–623.

bandwidth efficiency. These modulations require, however, high fidelity transmitters using highly linear amplifiers.

Unfortunately, linearity usually means very poor electrical efficiency, high cost, and low reliability. This problem has first been tackled by the engineers working on high capacity coaxial cable telephone systems, and in fact, most linearization techniques such as feedback, precorrection and feedforward, go back to the work done at Bell Labs some 50 years ago. However, techniques that are relatively simple to apply for low power amplifiers used in cable systems are usually much more difficult to apply to high power amplifiers such as those required for cellular radio. As a result, numerous engineers have stepped back from the practical difficulties until it became clear that the economical benefits from these techniques justified the effort to overcome these difficulties.

This book takes the reader through all the aspects of radio frequency power amplifiers that are necessary to understand the challenges and the particular behaviors of high power amplifiers linearized through the use of the feedforward technique. All the practical difficulties that have made a simple concept such as feedforward a difficult technique to apply are reviewed in detail together with their most recent solutions.

At the time where third generation radiocommunication systems are taking shape, this book will provide very valuable insights on the status of the technology to all engineers involved in the new systems' development. Those who will have to design, or to specify, such amplifiers will find the information necessary to make technical choices more knowingly and more rapidly.

Jack Powell
President, Telia SA (France)

Preface

The general subject of this book is linear amplification for radio frequency transmitters. Specifically, the book is concerned with the use of feedforward as a linearization technique for radio frequency power amplifiers and is primarily aimed at engineers and technicians but is suitable for anyone wishing to learn about linear amplifiers in general, and feedforward in particular. Although the emphasis is on one particular linearization technique, the general requirement for linearity in power amplifiers is discussed in detail relating to bandwidth, output power, efficiency, and distortion performance; a review of other linearization and synthesis techniques is also given.

With the evolution of existing and new standards for mobile communication systems and wireless multimedia services, the quantity and complexity of the signals to be transmitted from a single location is increasing. The demands placed on radio frequency power amplifiers, which are used in such systems, are subsequently increasing in terms of bandwidth, output power, efficiency, and allowable level of output distortion. There is a growing need for amplifiers, which amplify all types of signals without adding significant distortion and are capable of operating over a wide bandwidth and at potentially high levels of output power.

The primary goal of any radio system is to transmit and/or receive information. For broadcast radio and radio telephony, information is usually in the form of speech; however, text, pictures, and video are increasingly being used for data transfer and wireless multimedia applications. For example, most analog first generation cellular systems are limited to speech, but the development of second generation digital systems is allowing both speech and limited data capabilities. Future third generation systems and indeed developed second generation systems are expected to support much higher rates of data transfer, thus enabling wireless multimedia services such as video and the Internet to become a reality. One of the reasons that it is becoming more practical and cost effective to offer such services is that radio frequency power amplifiers, which are inherently nonlinear, can now be built to very high specifications and fulfill the requirements of a "linear amplifier." This has not been possible thus far using traditional techniques because the amplifiers generate distortion in the form of intermodulation and spectral regrowth. Power generated outside of the transmit channel causes interference in adjacent radio channels, while power generated in-band can cause errors in signal vectors and hence, a degradation in demodulation accuracy.

Because of their versatility and flexibility, linear amplifiers are finding an increasing number of applications in cellular radio systems, personal communication systems (PCS), international mobile telecommunications in the year 2000 systems (IMT-2000), and universal mobile telecommunications systems (UMTS). Linear amplifiers are capable of amplifying single-carrier and multicarrier signals, analog and digital signals, and constant envelope and nonconstant envelope signals. Linear amplifiers are thus effectively transparent to the modulation format and number of carriers. Furthermore, depending upon the choice of linearization technique, linear amplifiers can operate with low levels of distortion over the wide bandwidths that are necessary to support high data rate services such as the Internet and wireless multimedia. For example, wideband linear amplifiers are an integral part of the third generation system UTRA (UMTS terrestrial radio access), which is based upon wideband code division multiple access (WCDMA). The system currently has channel bandwidths of 5 MHz and data rates as high as 2 Mbps, enabling it to fully support UMTS/IMT-2000 requirements.

One of the primary goals of amplifier design is to produce an amplifier that has good efficiency and low distortion; however, in practice there is a trade-off between distortion performance and efficiency. For example, so-called Class A amplifiers have good distortion performance but low efficiency while so-called Class C amplifiers and to some extent Class B amplifiers are reasonably efficient but introduce significant distortion. As the power level increases, efficiency becomes more important. For example, the power level in multicarrier cellular radio systems can be of the order of 100W to 200W average with a corresponding peak power requirement of 1 kW to 2 kW. To minimize heat dissipation in the amplifier, it is therefore clearly desirable to have the best achievable efficiency for such applications; unfortunately this means using an amplifier that is nonlinear. Since linearity is very important for many applications, radio frequency power amplifiers are therefore operated in Class A with the consequent heat dissipation tolerated.

A number of techniques, referred to as linearization techniques, have been developed that eliminate or reduce the amount of distortion added by an inherently nonlinear power amplifier. In addition, methods have also been developed whereby a linear signal is generated using the synthesis of other nonlinear signals. The result is a linear radio frequency power amplifier. A commonly used linearization method for correcting distortion in amplifiers is to apply negative feedback, however such a technique is inherently bandwidth limited and is not suitable for wideband applications such as WCDMA. An alternative approach, which is suitable for wideband applications, is to use Class AB amplifiers, which are more efficient than Class A, and apply feedforward linearization. Improvements in distortion performance of 30 dB (×1000) can be achieved using feedforward in its basic configuration, although dual-loop feedforward or a combination of feedforward and another linearization method, such as predistortion, can yield even greater improvements.

In this book, Chapter 1 begins with an overview of feedforward and includes an introduction to a typical radio system for cellular, PCS, or IMT-2000/UMTS systems; some concepts specific to operation at radio frequencies are also discussed. In Chapter 2, a mathematical-based approach is used to discuss amplifier input-output characteristics, signal modulation formats, signal envelopes, peak-to-average ratios, and statistical analysis. Power amplifiers and system design are discussed in Chapter 3; topics include transistors for RF power amplifiers, amplifier efficiency, class of operation, intermodulation performance, and system design issues such as combining RF signals. The concept of a linear amplifier is introduced and practical examples are given for different system configurations. Chapter 4 reviews different linearization techniques including feedback (RF feedback, envelope feedback, Cartesian loop, and polar loop feedback), RF synthesis, envelope elimination and restoration, predistortion, and feedforward. The discussion on feedforward includes the principle of operation, signal cancellation, and loop control; dual-loop feedforward is also described. In Chapter 5, a detailed analysis of feedforward performance is given, with topics such as gain, input/output match, noise figure, broadband signal cancellation, error amplifier performance, and system efficiency. Throughout the text, practical examples are given; for example, frequency spectra and typical performance figures of commercial linear amplifiers.

Acknowledgments

I would like to express my gratitude to all those people who have helped me directly and indirectly with this book. In particular, I would like to thank Jack Powell and Jacques Rambaud of Telia SA (France) for their enthusiasm and support; they have contributed significantly to my understanding of amplifiers in general and linearization techniques in particular. A special thanks also goes to those people who have helped in the development and review of the manuscript, from its conception to completion: Wyc Slingsby, Ross Wilkinson, Kieran Parsons, Magnus Kvist, and Jack Powell. I would also like to thank my colleagues at Nokia, Ericsson, Radio Design (Sweden), the Defence Research Agency (UK), and the University of Bristol (UK) for their contribution to my knowledge and understanding of radio. Finally, I thank Toracomm Ltd. (UK), who provided considerable support throughout this project.

Further information on the products referred to in the text can be obtained from:

Toracomm Ltd.
toracomm@toracomm.co.uk
www.toracomm.co.uk
Tel. +44 117 900 8133

Telia SA (Allen Telecom)
jpowell@atlantic-line.fr
www.allentele.com
Tel. +33 556 89 5619

Ericsson Components
www.eka.ericsson.se
info@eka.ericsson.se
Tel. +46 8757 5000

1

Overview

1.1 Fundamentals—Radio Waves and Radio Frequencies

Radio waves and radio frequencies (RF) are terms used to identify a particular region of the much larger electromagnetic spectrum. Electromagnetics is itself defined as "the study of the electric and magnetic phenomena caused by electric charges at rest or in motion," and the fundamental physical quantities in electromagnetics are electric and magnetic fields.

Field theory can become very complicated and difficult to use, even for seemingly simple situations; however, certain approximations can be made in many cases to simplify calculations. For example, circuit theory, transmission line theory, and optical ray theory all have their origins in electromagnetic theory but are often much more practical and easier to use in certain well-defined situations. In circuit theory, for example, voltage and current, which are easier to measure and use in calculations, can be used instead of their electric and magnetic field counterparts. In some cases though, only field theory will give the correct explanations and answers. When to use one particular theory and not another is an important consideration and is closely related to the concept of electrical length discussed in Section 1.1.2. All electromagnetic waves obey the same fundamental laws, Maxwell's equations of electromagnetism, and are uniquely defined by their frequency, amplitude, and phase.

1.1.1 Frequency, Amplitude, and Phase

Frequency is measured in Hertz (Hz)—that is, the number of cycles per second—but angular frequency, which is usually given as the symbol ω—is also

1

widely used. The angular frequency, which describes the rate of change of signal phase with respect to time, is an important property of radio waves and can be calculated from

$$\omega = 2 \cdot \pi \cdot f \tag{1.1}$$

The radio part of the electromagnetic spectrum includes frequencies from the kilohertz range (10^3 Hz) up to the gigahertz range (10^9 Hz) with the cellular, PCS, and IMT2000/UMTS bands located between $\approx$400 MHz and 2.2 GHz.

Amplitude describes the magnitude of a signal (e.g., a voltage) and distinctions are usually made between instantaneous, average or root mean square (rms), and peak quantities. The average or rms value is equal to the instantaneous value averaged over one cycle (period T), that is,

$$v_{rms}^2 = \frac{1}{T} \cdot \int_0^T v^2(t)\, dt \tag{1.2}$$

The rms voltage is equivalent to the dc voltage, which would have the same heating effect as one cycle of the ac wave; for a sine wave the rms value is $1/\sqrt{2}$ or 0.707 times the peak value.

Phase is an angular quantity measured in degrees or radians (360 degrees or 2π radians per Hertz) and the rate of change of phase with respect to time is equal to the angular frequency. With respect to frequency, the rate of change of phase is equal to the delay, that is,

$$\tau = \frac{\Delta\phi}{\Delta\omega} \qquad \tau = \text{delay(sec)} \tag{1.3}$$

Delay in amplifiers and transmission lines is an important property of radio frequency circuits and is discussed in detail in later chapters. Note also that phase measurements are always relative to a defined reference.

1.1.2 Wavelength and Electrical Length

The wavelength λ is the distance a wave travels in one cycle, that is,

$$\lambda = \frac{\text{velocity}}{\text{frequency}} = \frac{v}{f} \tag{1.4}$$

λ = wavelength (m)
v = velocity (m/s), 3×10^8 m/s in air
f = frequency (Hz)

At low frequencies, such as < 60 Hz, wavelengths are extremely long (> 5000 km); while at high frequencies, such as visible-light, wavelengths are very short (10^{-6}m). For radio waves traveling through air, such as 300 MHz to 3 GHz (the UHF frequency band), wavelengths are in the 1m to 1 cm range.

In addition to being a function of frequency, wavelength also depends upon the material properties of the medium through which a wave is traveling since velocity is inversely proportional to the permittivity or dielectric constant ε_r of the medium. That is, (1.4) can be rewritten as

$$\lambda = \frac{c}{\sqrt{\varepsilon_r} \cdot f} \qquad (1.5)$$

λ = wavelength (m)
c = speed of light c (3×10^8 m/s)
f = frequency (Hz)
ε_r = relative permittivity or dielectric constant (dimensionless)

For example, a radio wave having a frequency of 1500 MHz and traveling in air has a velocity of 3×10^8 m/s and a wavelength of 20 cm ($\varepsilon_r = 1$). The same wave traveling on a printed circuit board with a dielectric constant $\varepsilon_r = 9$ travels more slowly at 1.0×10^8 m/s and has a shorter wavelength (6.7 cm).

The electrical length of an object (e.g., a component, piece of wire, cable, and cavity) is the ratio of its physical size compared to the wavelength at a particular frequency of interest and has important consequences in terms of an object's response to time varying fields such as radio waves. Electrical length is often expressed in terms of wavelength, such as 0.1λ and 2λ, but it can also be expressed in degrees (360 degrees = 1λ). For example, a piece of wire 10-cm long is electrically very small at low frequencies (a fraction of a wavelength) but electrically large (comparable to or greater than a wavelength) at radio frequencies. When an object is electrically small, there is little or no radiation and voltage and current concepts are valid; however, when an object is electrically large, radiation and other effects can occur and the electric and magnetic fields should be considered. Radiation due to poor shielding of components and unwanted signals traveling on inadequately grounded cables can become critical issues in radio frequency circuits since performance may be adversely affected.

1.1.3 Permittivity and Dielectric Loss

All materials can be classified on the basis of their electrical conductivity as conductors, semiconductors, or insulators. Good conducting materials such as copper, silver, or gold have a large number of electrons that are free to move about and "conduct" while dielectric materials such as air are electrically neutral (no free charge carriers) and are usually described by their permittivity. Permittivity is a dimensionless complex quantity; that is, it has no units and real and imaginary parts. The real part (the dielectric constant ε_r) relates to the ability of a material to *store* electrical energy while the imaginary part is a measure of how lossy a particular material is, that is, how much energy is converted to heat.

In the context of radio frequency design, important consequences of complex permittivity include:

- At a given frequency, circuits can be made smaller using materials with higher dielectric constants since the wavelength is shorter.

- The same radio wave traveling on different printed circuit board materials (e.g., Teflon with $\varepsilon_r = 2$ or Alumina $\varepsilon_r = 10$) has different wavelengths and this can affect the physical dimensions of a circuit, for example, a quarter wave, $\lambda/4$ coupler, or transformer.

- Loss, in any form, means conversion of useful energy in the form of radio waves to heat and should be accounted for. For example, lengths of tracks or cables can have significant loss that is easily "forgotten." Loss is particularly important when power levels are high, as is the case with power amplifiers, or very low, as in the case of low noise receivers.

The ratio of the imaginary and real parts of the complex permittivity, called the loss tangent ($\tan \delta$), is often used to describe the loss of a particular dielectric material. A "perfect" lossless material has a loss tangent of zero, but typical practical values for printed circuit board material are in the range $0.01 \rightarrow 0.0001$. Dielectric losses increase with frequency and are also temperature dependent, a factor that must be compensated for in certain applications, such as feedforward amplifiers.

1.1.4 Average and Peak Power

Power is proportional to the square of the voltage or electric field strength and has units of Watts (W). As with voltage or electric field strength it is usual to distinguish between instantaneous, average or rms, and peak quantities. However, unlike voltage or electric field strength, which are vector quantities, power is a scalar quantity having magnitude only.

The total power (or energy per second) in a given volume of space is equal to the sum of:

- Stored power;
- Dissipated power;
- Net power flow into the volume.

Stored power is energy contained in electric and magnetic fields (e.g., capacitors and inductors have stored electrical and magnetic energy), while dissipated power is energy in the form of heat due to conductor or dielectric losses. Power flow is energy being transported as a propagating electromagnetic wave, and it is this property that gives rise to the "effect at a distance" phenomena such as all forms of radio, radar, and satellite communication. Marconi's first demonstration of radio transmission across the Atlantic Ocean is a classic example whereas now it is becoming a part of everyday life for many people.

Average or rms power is calculated using the rms voltage given in (1.2) and has the same thermal effects as the instantaneous power over one cycle. Peak power is calculated using the peak (maximum instantaneous) voltage.

For a signal containing several frequency components, that is, an information-bearing signal (e.g., a modulated carrier or a multicarrier signal), the total average power, in terms of voltages, is proportional to the *sum of the voltage squares* whereas the peak power is equal to the *square of the voltage sum*. That is, for a resistance $R(\Omega)$,

$$P_{av} = \frac{\left(V_{1_{rms}}\right)^2 + \left(V_{2_{rms}}\right)^2 + \left(V_{3_{rms}}\right)^2 + \cdots}{R} \tag{1.6}$$

$$P_{peak} = \frac{\left(V_{1_{max}} + V_{2_{max}} + V_{3_{max}} + \cdots\right)^2}{R} \tag{1.7}$$

As shown in later chapters, calculation of the average and peak power and hence the peak-to-average ratio of single- and multicarrier modulated signals is an extremely important part of feedforward amplifier design.

1.1.5 Transmission Lines and Characteristic Impedance

Transmission lines consist of two or more conductors and are typically used to transfer power and/or information from one point to another; in some specialized applications, transmission lines are also used as circuit elements.

A complete analysis of any transmission line using Maxwell's equations reveals many different modes of wave propagation. Many of these modes, including so-called *waveguide modes*, can only propagate above a certain cut-off frequency that is determined by the geometry of the structure. If the distance between the conductors is small, as it typically is, then the cut-off frequency is very high and at lower radio frequencies the only mode of propagation is the so-called transverse electric and magnetic (TEM) mode. A TEM mode is characterized by the electric and magnetic fields only having components transverse to the direction of propagation and power flow. A special property of the TEM mode is that voltage and current concepts are valid since the dependence of time-varying field components in the transverse direction is the same as for the static (non-time-varying) situation.

The *characteristic impedance* of a transmission line, often has units of Ω and is given the symbol Z_0, is equal to the ratio of the voltage and current wave amplitudes and can be calculated from

$$Z_0 = \sqrt{\frac{R + j \cdot \omega \cdot L}{G + j \cdot \omega \cdot C}} \qquad (1.8)$$

R = Series resistance per unit length (Ω/m)
L = Series inductance per unit length (H/m)
G = Shunt conductance per unit length (S/m)
C = Shunt capacitance per unit length (F/M)

The characteristic impedance of a transmission line is thus dependent only on the distributed parameters R, L, G, and C and is independent of its length. Transmission lines can therefore be made to have a specific characteristic impedance, such as 50Ω, by varying the physical construction of the line and hence the value of the distributed parameters.

1.1.6 Impedance Matching, VSWR, and Return Loss

In circuits where signals do not vary with respect to time, that is, dc circuits, maximum power is transferred from a source to a load if the load resistance R_L equals the source resistance R_s. When dealing with time-varying signals such as radio waves, resistance or rather *impedance* matching is also necessary to ensure the efficient transfer of power from a source to a load. That is, if the source impedance is not matched to the load impedance ($Z_L \neq Z_S$), then energy is reflected, causing a reduction of power in the forward direction. Furthermore, impedance mismatching also results in the formation of standing waves; the sum of incident and reflected voltages. Standing waves due to impedance mismatch can cause frequency-dependent amplitude ripple, an undesirable property in many applications.

The ratio of incident and reflected signals due to an impedance mismatch, $Z_L \neq Z_S$, is called the *voltage reflection coefficient* Γ and is calculated from

$$\Gamma = \frac{Z_L - Z_S}{Z_L + Z_S} \tag{1.9}$$

The voltage reflection coefficient is, in general, a complex quantity since the impedances Z_L and/or Z_S can contain inductive or capacitive elements, that is, reactance as well as resistance. A perfect match and maximum power transfer is obtained when the load impedance is equal to the complex conjugate of the source impedance, that is, any source reactance is resonated with an equal and opposite load reactance, leaving only equal resistance values and hence no reflection. In the case of a transmission line, the line is said to be matched when the source and load impedances are equal to the characteristic impedance of the line, that is, $Z_L = Z_0 = Z_S$.

The reflected power is calculated using the *magnitude* of the voltage reflection coefficient and when expressed in decibels, that is, $-20 \log(|\Gamma|)$, the reflected power is referred to as the *return loss*.

An alternative method of specifying reflected power is to use the voltage standing wave ratio (VSWR), which describes the ratio of the maximum and minimum voltages. The relationship between return loss and VSWR is

$$S = \frac{1 + |\Gamma|}{1 - |\Gamma|} \tag{1.10}$$

A perfect match occurs when the source impedance is equal to the load impedance and $|\Gamma| = 0$; the return loss is then equal to $-\infty$ dB and the VSWR,

$S = 1$. A return loss of <20 dB ($|\Gamma| = 0.1$) or VSWR of <1.2 is considered to be a very good match since the power in the reflected signal is <1/100 of the incident power. For example, if the incident power is 10W, then <0.1W is lost (reflected) due to the impedance mismatch. The system link budget (Chapter 3)—which specifies gains, losses, and power levels—can be used to specify the maximum allowable impedance mismatch at different points within a system.

In RF circuits, transmission line structures such as coaxial cables, microstrip, and stripline that have a resistive or real characteristic impedance (typically 50Ω) are used for signal connection between different circuit components. However, since the input and output impedances of these components (often amplifiers) nearly always have some capacitive or inductive element and a different resistive component, special impedance matching circuits are required.

Simple impedance matching circuits are only applicable for narrow bandwidths; in systems that require good impedance match over a wide bandwidth (e.g., an octave), broadband matching techniques such as multisection "pi" or "T" networks, ferrite loaded transmission line transformers, multisection quarter wave transformers, or Quadrature hybrids are used.

1.1.7 Units—dB, dBm, and dBc

In principle, the range of possible voltage and power levels is enormous; in RF circuits for example, the power level can vary from attowatts 10^{-18}W to kilowatts 10^3W. A large linear range can be compressed to something much more manageable if a logarithmic scale rather than a linear scale is used. Logarithms also have several other advantageous mathematical properties such as logarithmic addition replacing linear multiplication and are frequently used when calculating gain and power levels in RF circuits. There are also historical reasons, mostly originating from the early days of telephony, for using logarithmic scales; for example, the human ear's response to sound is on a logarithmic scale.

The decibel scale (dB), which is a logarithmic scale, represents the logarithmic ratio between *any* two power levels. Taking the load and source powers P_L and P_S as examples, the decibel scale is defined as

$$dB = 10 \cdot \log_{10}\left(\frac{P_L}{P_S}\right) \qquad (1.11)$$

where P_L and P_S are measured in Watts. Rewriting (1.11) in terms of voltage and impedance rather than power gives

$$dB = 10 \cdot \log_{10}\left[\frac{\left(\dfrac{V_L^2}{Z_L}\right)}{\left(\dfrac{V_S^2}{Z_S}\right)}\right] \qquad (1.12)$$

where
Z_L = load impedance (Ω)
Z_S = source impedance (Ω)

At constant impedance, that is, $Z_L = Z_S$ (50Ω in most radio equipment) (1.12) simplifies to

$$dB = 20 \cdot \log_{10}\left(\frac{V_L}{V_S}\right) \qquad (1.13)$$

Thus, regardless of whether power or voltage is used, the answer in decibels is the same provided that the impedance is constant; it is only the linear gains that are different. For example, a linear power gain of 10 and a linear voltage gain of 3.16 are both equal to 10 dB at constant impedance. Negative gain, that is, a loss, occurs when the voltage or power ratio is less than unity.

Note also that decibels are a dimensionless scalar quantity and that there must always be a reference level, either a standard reference level such as 1 mW or an arbitrary one, for example the signal level at some specified point in a circuit.

Decibels are simply the ratio of two numbers and give no information about the absolute level of a signal; for example, 10W compared to 1W has the same decibel ratio as 100 mW compared to 10 mW. For absolute levels, a different scale (dBm) is used where the reference level is fixed at 1 mW. Table 1.1 shows some examples of power levels on a dBm scale.

Most measuring instruments (power meters, network and spectrum analyzers, and signal generators) typically have a choice of display format; however, the important factor is to remember whether the measurement is relative, for example a gain, or absolute, that is, the power in Watts.

One other logarithmic scale of measurement used with power amplifiers is dBc, where the reference level is a carrier wave. Like ordinary decibels,

Table 1.1
The dBm Scale for Absolute Power

Power (dBm)	Power	Example
−174	4×10^{-21}W	Thermal noise (1-Hz bandwidth, room temperature)
−110	1×10^{-14}W	Receiver sensitivity level example
0	1 mW	Input power to a power amplifier
30	1W	Power amplifier output (low power)
50	100W	Power amplifier output (high power)

dBc is a relative measurement; for example, −30 dBc means a signal 1,000 times smaller than the reference carrier.

1.2 System Overview

By definition, radio systems use radio waves to transmit and/or receive information. Figure 1.1 shows the basic building blocks of a typical mobile radio system. Essentially, there are three distinct parts:

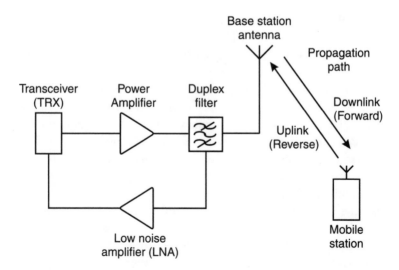

Figure 1.1 Cellular radio system block diagram (single channel).

- The radio base station (RBS or simply BS);
- The propagation path;
- The mobile station (MS).

In principle, both the BS and the MS can act simultaneously as a transmitter and receiver, that is, full duplex communication is possible; however, some systems support only simplex communication.

1.2.1 Transceivers

As the name implies, a transceiver, commonly referred to as a TRX, acts as a transmitter and a receiver; its primary function when transmitting is *modulation* and *up-conversion* and when receiving, *down-conversion* and *demodulation*. Note that a TRX also has many other functions, such as channel coding, encryption, and power regulation.

Modulation is the process whereby information is added to a pure sine wave known as a carrier wave or CW signal. The most common forms of modulation—that is, amplitude, phase, and frequency modulation (AM, PM, and FM, respectively)—involve altering the amplitude, phase, and/or frequency of a carrier wave in a predetermined manner. The process of demodulation is simply the reverse; that is, information is recovered in its original form. A large number of analog and digital modulation schemes exist with the latter becoming increasingly and widely used.

Up-conversion is the process of increasing the frequency of a signal, for example from baseband to RF, and is often done in several stages using intermediate frequencies (IF). The baseband information (for coded speech, a signal with frequency content typically <8 kHz) is first modulated onto an IF frequency (in general <100 MHz) before being up-converted to RF. The frequency of the RF carrier wave, which usually has a much higher frequency than the modulating or IF signal, is determined by the specific part of the spectrum on which a particular network operator has permission to transmit. For example, a PDC operator is currently allocated a block of spectrum in the 800- or 1500-MHz band; a GSM operator in the 900-, 1800-, or 1900-MHz band; and an NMT operator in the 450- or 900-MHz band. The precise frequency within these blocks of allocated spectrum is chosen on the basis of local factors, that is, system capacity and coverage issues.

The modulation and up-conversion process, which is done at low power, has the effect of transferring the baseband data up to RF without loss of information content. Down-conversion and demodulation are simply the opposite.

Traditionally, each TRX has its own power amplifier stage and can thus generate a single high-power modulated carrier anywhere in the transmit band. For example, in standard NMT the output power rating of a TRX is 50W, which can be considered high power. Thus, for a radio system using 20 TRXs there are 20 high-power amplifiers; as the number of TRXs increases, so too does the number of power amplifiers.

New TRX designs are currently being developed and offer many potential advantages, including:

- Multicarrier signal generation;
- The ability to support more than one radio standard;
- Low output power, thus removing the need for a power amplifier stage.

Advances in digital signal processing (DSP) techniques are enabling the development of such "multistandard" or "software-configured" radio systems, but there is also a parallel development in amplifier hardware architectures as these systems rely heavily on linear power amplifier technology.

1.2.2 Voltage and Power Amplifiers

An amplifier is typically characterized by its gain, linearity, noise performance, and input and output match. The function or purpose of an amplifier is to produce a signal at the output that is identical in all respects to the signal at the input; the amplifier is then said to be distortion-free. Note that although most amplifiers in their basic form perform the process of linear multiplication, gain is usually quoted in decibels assuming matched source and load impedances.

Depending upon the value of the source and load impedances with which the amplifier is designed to work, an amplifier can be described as either a *voltage amplifier* or a *power amplifier*. A voltage amplifier, for example, is designed to work with and therefore is characterized with high impedances (kΩ, MΩ). Although the voltage gain may be high, the current drawn from the source and the current supplied to the load are relatively small due to the high impedances. A preamplifier in a home stereo is an example of a voltage amplifier. Alternatively, power amplifiers are used and characterized with much lower impedances (typically 50Ω) and are capable of delivering large amounts of power to a load.

1.2.3 Signal Amplification and Linearity

An ideal amplifier does not produce any distortion, that is, no unwanted changes in the time domain waveform or frequency spectrum of the input signal; the output is related to the input in a purely linear fashion. In practice, however, the components used in amplifiers, such as transistors, have non-linearities that result in distortion of the output signal. The degree to which a particular component is nonlinear depends to a large extent on the signal level and biasing arrangement and is most often a complex relationship. Chapters 2 and 3 discuss nonlinear behavior and bias techniques in detail.

A simple appreciation of the need for linearity can be gained by considering the power amplifier stage that drives the loudspeakers in a music system. The goal is to reproduce both soft and loud passages of music without distortion, and this only happens if the amplifier is linear. Power amplifiers for radio systems have a similar requirement, although the definition of acceptable distortion is more complex.

1.2.4 Power Transistors

The most common device for producing power at RF is the transistor. Traveling wave tubes (TWTs), which are traditionally used at higher microwave frequencies, are gradually becoming available for RF frequencies; however, the basic building block of most RF power amplifiers is currently a power transistor.

There is no single optimum technology for making transistors, rather a collection of processes that have advantages and disadvantages depending primarily upon the desired frequency of operation, output power level, and linearity. In general, the higher the frequency and the higher the output power, the more difficult it is to make transistors with the required performance. Operation at higher frequency is more difficult because effects such as internal parasitic capacitances and the length of bond wires become more significant (the wavelength is smaller) while operation at higher power means that more heat is dissipated, thus increasing the thermal requirements for safe and reliable operation.

The internal parasitic capacitances within a transistor result in a typical gain and output power reduction of 6 dB per octave. Transistors designed for operation at higher frequencies use smaller feature sizes to reduce the parasitics; however, this reduces the current-carrying capacity of the signal tracks within the device and, furthermore, it becomes more difficult to extract unwanted heat from the smaller die. The result is reduced output power capability.

Most RF power transistors do not consist of a single transistor device but are constructed from paralleled blocks of die. This paralleling can be at semiconductor level, for example, interdigitated transistors, or at package level, where multiple blocks of die are connected in parallel by bond wires (devices have been constructed with 20 or more die in parallel). The complexity inside the package is further increased when internal matching is used. The bond wires are used as series inductors and capacitor die are used as shunt capacitors to form a simple low-pass network that is used to transform the relatively low input and output impedances to higher values. Since the networks are low-loss (gold bond wires, silicon nitride capacitors) and physically close to the semiconductor, high-Q and low-loss, wideband matching is possible. There is, however, a limitation to the number of die that can be combined before other parasitic effects and potential instabilities occur. Since there is some flexibility in the height and length of the bond wires, it is possible to compensate for batch-to-batch variations in die characteristics by varying the bond wire settings on the assembly line. In some applications such as very high power pulse devices, double input and output matching is used.

Another method that is commonly used to increase the power from a single device is the push-pull or "Gemini" package. In this, two separate "transistors" (which may themselves be constructed from internally matched, paralleled, interdigitated transistors) are mounted on a single flange. The "half-transistors" are driven 180 degrees out-of-phase and the respective output signals are then recombined with another 180-degree combiner. This technique has several advantages, namely:

- The power is shared between two transistors.
- The impedances at the input and outputs of the 180-degree splitter and combiner are four times higher than the individual device impedances, reducing the transformation ratio of matching circuits (thus reducing loss and improving bandwidth).
- The physical arrangement of the input and output terminals of the Gemini package minimizes the parasitic series inductance when coaxial transmission line transformers are used.
- Balanced operation minimizes circulating ground currents and leads to even harmonic cancellation.

Currently for high power at RF (up to about 2.5 GHz), the choice of transistor technology is effectively between:

- Silicon bipolar transistors;
- Gallium Arsenide Metal Semi-Conductor Field Effect Transistors (GaAs MESFET);
- Silicon Metal Oxide Semiconductor (MOS) transistors; for example, Laterally-Diffused MOS (LDMOS).

Table 1.2 shows some very basic transistor parameters, however, the performance of power amplifiers in terms of choice of transistor, class of operation, linearity, efficiency, and power supply requirements is delayed until later chapters.

It is not possible to make absolute statements about which transistor technology is "optimum" because there are so many factors to consider and the requirements can vary considerably between applications, however, some very general points are:

- Silicon devices can be operated at higher supply voltages than their Gallium Arsenide counterparts thus enabling them to draw less current for a given power (silicon devices that can be used with 48-V supply voltage are also being developed).

- The higher electron mobility of Gallium Arsenide compared to Silicon makes it better suited for operation at higher frequencies.

- Gallium Arsenide technology is very well established while technologies such as LDMOS are relatively new.

Table 1.2
RF Power Transistors

Transistor	Material	Input	Control parameter	Supply voltage
Bipolar	Silicon	Semiconductor junction	Current	Typ. +26V
MESFET	GaAs	Semiconductor junction	Voltage	Typ. +12V
MOSFET e.g. LDMOS	Silicon	Insulating layer	Voltage	Typ. +26V

- LDMOS and other FET families such as VFET and ISOFET generally offer simplified biasing, higher gain, and improved ruggedness into mismatches.

As Figure 1.2 shows, for high power levels—that is, currents in the ampere range and power dissipation of tens of watts—the physical structure, packaging, and specification of high-power transistors makes them different from their low-power counterparts. In addition to producing a high-power RF signal, power transistors also generate large amounts of heat, causing a rise in the junction temperature of the semiconductor materials; self-heating also causes the ambient temperature inside the equipment to increase. Thermal considerations are thus extremely important in power amplifier design, and heat must be transported away from transistors as efficiently as possible. Great care should be taken to ensure that even at maximum ambient temperature, which can be as high as 50° to 55°C, internal power dissipation and self-heating does not degrade performance or cause permanent damage to components.

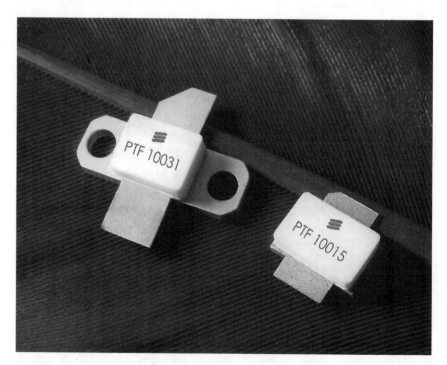

Figure 1.2   (a) LDMOS power transistors for large signal applications up to 1 GHz. (Courtesy of Ericsson Components AB).

Figure 1.2 (continued) (b) Ericsson 70W LDMOS power transistor for use in DAMPS and GSM (800–900 MHz band).

For example, to avoid damaging transistors, the junction temperature should not be allowed to exceed some maximum value—typically 200°C for silicon RF power devices. A typical design maximum for the junction temperature during operation in a BS should be 150°C, and a general rule is that device lifetime is halved for each 10° increase in junction temperature. In regard to MTBF (Mean Time Between Failure), a statistical analysis of a system with many combined devices, such as a BS with a design life of 10 to 15 years, indicates that a device MTBF well in excess of 100 years is required if frequent replacement of individual devices is to be avoided.

1.2.5 Directional Couplers

Directional couplers are frequently used in RF circuits as precision samplers (power splitters) and power combiners with directional characteristics to ensure that the coupling is unaffected by reflected or reverse injected signals. For example, in feedforward amplifiers, directional couplers act as power splitters at the feedforward input and the output of the main amplifier and as power

Figure 1.2 (continued) (c) Ericsson Silicon bipolar power transistors for wideband CDMA applications at 2.1–2.2 GHz.

combiners for distortion and carrier cancellation (Chapter 4). Important parameters of directional couplers include insertion loss, coupling coefficient, isolation, directivity, power handling, and bandwidth of operation.

1.2.5.1 Insertion Loss and Coupling Coefficient

Figure 1.3(a) shows the voltage distribution in a coupler with input voltage V_{in} and output voltage V_{out}. The ratio between output and input voltages, that is V_{out}/V_{in}, is given by the symbol α and when expressed in decibels is equal to the coupler insertion loss A (Figure 1.3b), that is,

$$A = 20 \cdot \log_{10}(\alpha) \tag{1.14}$$

The ratio of the voltage at the coupled port and the input port, that is V_c/V_{in}, is called the voltage coupling coefficient c. When expressed in decibels, the power coupling coefficient C is given by

$$C = 20 \cdot \log_{10}(c) \tag{1.15}$$

Figure 1.2 (continued) (d) Power transistors for wideband CDMA application.

If internal dielectric and conductor losses of the coupler are assumed negligible, then applying the law of conservation of power, the input power P_{in} is equal to the sum of the power at the other three ports, that is,

$$P_{in} = P_{out} + P_c + P_{iso} \qquad (1.16)$$

If perfect isolation between the input and isolated ports is assumed, that is, zero power at the isolated port ($V_{iso} = 0$, $P_{iso} = 0$), and all the ports are terminated with the same impedance R, then (1.16) becomes

$$\frac{V_{in}^2}{R} = \alpha^2 \cdot \frac{V_{in}^2}{R} + c^2 \cdot \frac{V_{in}^2}{R} \qquad (1.17)$$

Re-arranging gives

$$\alpha = \sqrt{1 - c^2} \qquad (1.18)$$

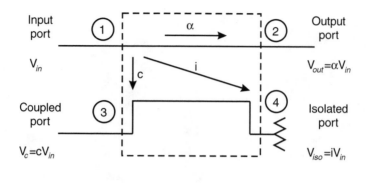

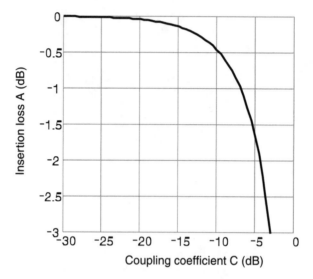

Figure 1.3 Directional coupler insertion loss and voltage distribution.

For example, assuming perfect isolation, a 10-dB coupler has the following parameters:

Power coupling coefficient	$C = -10$ dB
Voltage coupling coefficient	$c = 0.316$
Output voltage coefficient	$\alpha = 0.949$
Insertion loss	$A = -0.46$ dB

In practice, the insertion loss is higher due to internal dielectric and conductor losses and, furthermore, the isolation is finite, that is, power is dissipated at the isolated port ($P_{iso} \neq 0$).

1.2.5.2 Isolation and Directivity

Coupler isolation I (dB) and voltage ratio $I = V_{iso} / V_{in}$ are related by

$$I = 20 \cdot \log_{10}(i) \qquad (1.19)$$

For a perfect coupler the isolation is infinite, that is $i = 0$ and $I = -\infty$ dB. In practice, however, the isolation between the input and isolated ports of a coupler is finite, that is $i \neq 0$, $V_{iso} \neq 0$, and hence $P_{iso} \neq 0$. For example, if $i = 0.0316$, then the isolation $I = -30$ dB and the power appearing at the isolated port, that is $P_{iso} = i_2 V_{in}^{2} / R$, is dissipated as heat in the load resistor (typically $R = 50\Omega$).

Now consider a signal *incident* at the *output* of the coupler shown in Figure 1.3. Since a coupler is, in general, a symmetrical device, the output port (port 2) becomes the input port, the isolated port (port 4) becomes the coupled port, and the coupled port (port 3) becomes the isolated port. Port definition is thus a function of the direction of the input signal (forward or reverse). When both forward and reverse signals are present in a coupler, the property of *directivity* is used to describe the relative levels of forward and reverse signals.

Directivity, D (dB), is defined as

$$D = I - C \qquad (1.20)$$

That is, the directivity of a coupler is equal to the difference between the isolation and the coupling factor (dB). For example, when $C = -10$ dB and $I = -30$ dB, the directivity $D = -20$ dB. When the forward power at port one is for example 0 dBm, the coupled power at port two is -10 dBm. If a reverse signal, also of power 0 dBm, and incident at port two is present, the reverse signal level at port three (now the isolated port) is -30 dBm, that is, 20 dB lower than the forward signal. The coupler is thus said to be directional; that is, the forward signal level at port three is higher (by an amount equal to the directivity) than the reverse signal. As shown in Section 5.2, the coupler directivity is an important factor in determining the input and output return loss of a feedforward amplifier.

1.2.6 Duplex Filters

The final stage in a radio transmitter before the antenna is a filter that acts as a mask (an attenuator) for unwanted signals while allowing the desired signals through with minimum loss. The name duplex arises from the fact that it has a similar function when receiving and transmitting. When different frequency bands are used for receiving and transmitting, that is, frequency division duplex (FDD), the distance between them is referred to as the *duplex distance*. For example, the duplex distance in NMT is 10 MHz (450-MHz band) and in PDC the duplex distance is 48 MHz (1500-MHz band). Another mode of operation, referred to as time division duplex (TDD), is to use the same frequency band for transmitting and receiving.

Filters are passive devices and a final-stage duplex filter in a RBS typically has a transmit loss of less than 2 dB (for a MS the loss is typically 3 dB). The loss is primarily dependent on the type of filter and the number of network poles in the transfer function (the order of the filter). In general, a better frequency response is achieved with a higher number of poles, but the insertion loss is higher. Note, however, that even a high-order filter has only a limited ability to suppress undesired or spurious signals created by nonlinearities in devices such as transistors and mixers. Furthermore, filtering cannot remove unwanted signals that appear within or very close to the transmitter band since such filtering would also suppress the desired signals. Methods other than filtering therefore need to be used to ensure that requirements laid down in individual radio standard specifications are fulfilled.

1.2.7 Cables

As previously mentioned, a piece of wire longer than a few centimeters in length is electrically large at RF and therefore behaves as a *transmission line* (Section 1.1.5). To prevent unwanted radiation and loss of power due to impedance mismatches, all cables should be chosen to have good shielding properties and a characteristic impedance equal to the impedance of the signal source and load (e.g., double screened braid or solid copper coaxial cable with $Z_0 = 50\Omega$).

Like a filter, a coaxial cable or any other transmission line is a passive device and has loss. For example, it is not uncommon for losses between indoor and outdoor equipment, where the distance can typically be tens of meters, to be greater than 3 dB; that is, more than half the power is converted into heat. Using very large diameter cables loss can be minimized, but such cables are expensive and physically difficult to handle.

1.2.8 The Antenna

The function of an antenna is to radiate and/or receive energy in the form of traveling electromagnetic waves. An antenna that transmits and receives energy equally in all directions is called an *isotropic* radiator, while an antenna that has a favored direction of transmitting or receiving is referred to as a *directional* antenna. For example, a BS antenna is often designed to transmit/receive preferentially in a certain direction, that is, three antennas each covering 120 degrees instead of one covering 360 degrees. For an MS the opposite is true: the antenna is close to being omnidirectional (0 dBi) since it must have the ability to receive and transmit in all directions. An array antenna, that is, an antenna with many elements, may have a gain in excess of 20 dBi.

The majority of existing antennas for mobile radio systems use a single or a small number of elements; although some systems, such as those using adaptive antennas, are beginning to exploit the potential advantages of using an array. Note that due to sidelobes in the antenna radiation pattern, even a highly directional antenna still receives/transmits in directions other than desired. In most radar applications and in mobile radio systems that use narrow antenna beams to increase capacity and reduce interference, it is extremely important to minimize sidelobe levels.

An important system parameter is the effective radiated power, which is the power supplied to the antenna terminals after any losses in the feeder cable, multiplied by the gain of the antenna. For example, if the transmitter has an output power of 20W (43 dBm), the feeder has a loss of 3 dB, and the antenna gain is 9 dBi, then the effective isotropic radiated power (EIRP) is equal to 49 dBm, almost 80W. The "i" refers to an ideal isotropic antenna that radiates equally in all directions. EIRP is a useful way of comparing antennas with different gains and is frequently used in system analysis calculations such as the link budget (Chapter 3).

1.2.9 Propagation Path

Energy radiated from an antenna propagates through the atmosphere as a traveling electromagnetic wave in much the same way as visible light; that is, it is attenuated with distance and can be reflected, refracted, and diffracted.

In addition to free-space attenuation, a radio signal in the mobile environment is also subject to the effects of fading and shadowing and the resulting total attenuation can be of the order of 150 dB, that is, transmit 1W (30 dBm) and receive 1 FW $10^{-15}\Omega$ (−120 dBm). Such large attenuations place high demands on receivers and transmitters in both BSs and MSs. A receiver must be able to reliably detect signals having very little energy and a transmitter must be powerful enough to ensure that the minimum receiver level for reliable detection, that is, the receiver sensitivity, is achieved even under worst case path loss conditions.

1.2.10 Mobile Station

The requirements on an MS in a radio system differ quite significantly from those on the BS; a detailed discussion of the development and technology of mobile units is beyond the scope of this text. In relation to power amplifiers however, the following points can be made:

- MSs also require a power amplifier stage.
- The amplifier should be small, have low cost, and be as power efficient as possible to conserve battery life.
- The output power should be as low as possible to minimize power consumption.
- Normally the amplifier only has to amplify a single carrier.
- A single carrier can cause linearity problems if the signal envelope has amplitude variations (see Chapter 2).

1.2.11 Low-Noise Amplifiers

Unlike the previously mentioned power amplifiers, low noise amplifiers are specifically designed to cope with potentially very weak receive signals and the emphasis is on noise performance rather than power handling capability. The received signal-to-noise ratio (SNR) can be very low and thus, amplifiers that add as little additional noise of their own as possible, that is, have a low noise figure, are required. Traditionally, narrowband amplifiers are used in low noise amplifiers (LNAs), however, for wideband applications, a good candidate is a linear feedforward amplifier.

1.2.12 Access Techniques

Access techniques are the methods used to distinguish between users in a radio system and the most common techniques are time division multiple access (TDMA), frequency division multiple access (FDMA), and increasingly, code division multiple access (CDMA). For example, first generation cellular systems use FDMA while most second generation systems use TDMA together with FDMA. Future third generation systems are likely to use CDMA and possibly TDMA/FDMA variants of second generation systems.

FDMA is the traditional technique for radio broadcasting whereby the frequency band is divided into a number of discrete carrier frequencies. Each user is assigned a specific carrier frequency that is not shared with any other user; for example, a 10-MHz frequency band can be divided into 400 channels each having a 25-kHz bandwidth.

Alternatively, as the name suggests, TDMA involves separating users in time and enables several users to share the same carrier frequency—the information is sent in timeslots, one user followed by another. Most second generation radio systems use a combination of FDMA and TDMA whereby a user is first assigned a carrier frequency and then a timeslot. For example, in the cellular PDC standard there are six timeslots per carrier frequency and the duration of one timeslot is 667 μs; in GSM the corresponding numbers are eight and 577 μs.

When CDMA is used as the access technique, all users transmit on the same carrier frequency and at the same time. Instead of identifying a particular user by a particular carrier frequency or timeslot, each user is assigned a *unique code* that the receiver recognizes as belonging to that particular user. CDMA is the access technique used for second generation systems such as IS95 and is a strong candidate for third generation standards (IMT2000 and UMTS).

A comparison of the relative spectral efficiency, performance in fading environments, and overall system complexity for different access techniques is not required here; however, as will be discussed, the choice of access technique and associated modulation scheme does have important implications on the design of power amplifiers.

1.3 Feedforward Components

A detailed description of feedforward is given in Chapters 4 and 5; a brief overview is given here.

Feedforward design can be divided into the following general areas:

- Radio design;
- Mechanical design;
- Cooling/thermal design;
- Control system/DSP design.

The emphasis in the following sections is on radio design and performance; however in relation to the development of ultralinear power amplifiers, the importance of good mechanical and thermal design cannot be understated. The combination of high output power and low efficiency together with the requirements for good shielding, grounding, and high levels of isolation impose very tough demands on the mechanical and thermal design of such amplifiers.

1.3.1 Mechanical Design

In its basic form, a feedforward amplifier consists of two amplifiers (the main and error amplifiers), directional couplers, delay lines, and loop control networks (Figure 1.4). The main amplifier generates a *high*-power (distorted) output signal while the error amplifier produces a *low*-power distortion-cancellation signal. Directional couplers are typically used as power splitters and power combiners, and to ensure operation over a wide bandwidth signal, paths are matched using delay lines. Loop control networks, which adjust signal amplitude and phase, are used to maintain signal and distortion cancellation within the various feedforward loops. Associated with all these "components," however, is the *mechanics*—that is, heatsinks, chassis, shielding walls, covers, baseplates, fans, cable fixings, gaskets, and screws. Although the intention here is not to discuss, for example, different heatsink designs or grounding methods, it is important to realize that such a design is crucial to the overall success in terms of radio performance.

Thermal and mechanical design is not just about ensuring radio performance; for example, from a practical viewpoint, physical size, weight,

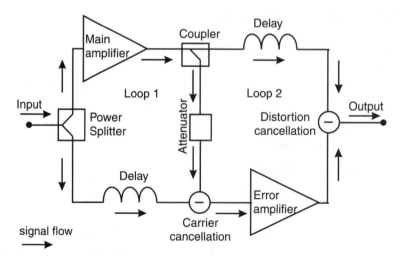

Figure 1.4 Feedforward components.

acoustic noise, and even appearance are all important factors in BS design. The current trend is towards more compact but higher performance BSs; and power amplifier stages can account for a significant proportion of the total size, weight, acoustic noise, and even the cost of a BS. It is therefore extremely important that mechanical and thermal considerations are not "forgotten" or given a lower priority than they really deserve. For example, power amplifiers are capable of radiating high levels of electromagnetic energy that may interfere with nearby equipment, and issues such as grounding and shielding should therefore be addressed at an early stage in the design process. In addition to such electromagnetic compatibility (EMC) issues, there are also environmental issues to consider. For example, components containing toxic substances are sometimes used in amplifiers. Investigation into alternative and more "environmentally friendly" components and manufacturing processes is one aspect of the overall design process.

1.3.2 Main Amplifier

The starting point for a feedforward amplifier is the main amplifier. As discussed in Chapter 3, amplifiers can be broadly compared in terms of linearity, efficiency, and power handling capability and the goal is then to have a suitable combination of linearity, output power, and efficiency for a specific application. It is also desirable that the amplifier has stable and well-defined characteristics and is capable of being manufactured in large quantities, that is, the minimum amount of manual trimming and production time. Manufacturing cost is another important factor and is primarily determined by the cost of the transistors and the mechanics; the power supply unit may also add significantly to the cost. Amplifiers, which require a relatively low number of transistors and simple mechanics (e.g., light and easy-to-machine heatsinks), are preferred to those that use many transistors and have complicated (heavy and difficult-to-machine) mechanics.

1.3.2.1 Average Output Power

In a feedforward system, the output power of the main amplifier largely determines the cooling requirements for the whole feedforward amplifier. In general, the efficiency of the main amplifier is low ($\approx 15\%$ for Class AB amplifiers; see Chapter 3); therefore, as the required output power increases, the amount of heat dissipation increases rapidly. To avoid overheating, improved cooling is then required—for example, a larger heatsink and hence a bigger and heavier amplifier—and/or an increased flow of air and more acoustic noise. In a feedforward system, it is therefore extremely desirable to minimize losses after the amplifier and, above all, to have a system configuration that requires low output power.

As described in Chapter 3, the average output power of a feedforward amplifier is determined by the system link budget and important parameters include losses between the amplifier and the antenna, antenna gain, path loss, and MS sensitivity. High output power is required if the amplifier(s) is physically remote from the antenna since the losses, mainly due to cables, are high. To overcome such loss, several amplifiers can be combined in parallel, but this in turn increases the power consumption, size, and cost of the final "amplifier unit." Typically, the average output power of a single feedforward amplifier in this type of configuration is in the 10- to 100-W range. Alternatively, the output power can be reduced significantly if the amplifiers are mounted close to the antenna (Section 3.11); typical feedforward output powers are then in the 1- to 5-W range.

In a feedforward system, the highest output power actually occurs at the output of the main amplifier and, therefore, losses between the amplifier and the antenna are often separated into internal and external losses; that is:

- Losses external to the feedforward amplifier such as loss of combining networks, cables, protection devices, test and measurement couplers, and filters;
- Losses internal to the feedforward amplifier such as coupler insertion losses and delay line loss.

External losses set the level of feedforward power, while internal losses set the level for the main amplifier.

As will be shown in Chapter 5, the output power of the main amplifier is higher than the feedforward output power due to internal losses, namely a delay line and directional couplers. Typically the total loss is ≈2 dB to 3 dB depending upon the specific application and choice of delay element. For example, if the couplers have coupling coefficients of −20 dB and −10 dB, the theoretical insertion loss due to the couplers is 0.5 dB (Figure 1.3); in practice, the loss is higher (typically ≈1 dB) due to internal dielectric and conductor losses. A delay line loss of ≈1 dB is obtained if the required delay is 12 ns and the delay element is a semirigid coaxial cable, which has a loss of 0.085 dB/ns (e.g., RG140). Note that for a fixed delay at a given frequency, the delay line loss can differ considerably depending upon the type of delay structure (Section 1.3.5). Note also that in dual-loop feedforward systems (Chapter 4), the output losses are much higher due to the inclusion of a second set of output couplers and delay line. In all cases, for a fixed feedforward output power, as the frequency or the amount of delay increases so too does the loss; hence, the output power of the main amplifier must be increased to compensate.

1.3.2.2 Peak Power

As discussed in Chapter 2, multicarrier signals typically have a high peak-to-average or peak-to-mean ratio ΔP_s. The peak power capability of the main amplifier, P_{1M}, must therefore be sufficiently high compared to the average power, P_M, to avoid distortion due to signal clipping on the peaks of the input signal; that is

$$\frac{P_{1M}}{P_M} > \Delta P_S \qquad (1.21)$$

It is possible to reduce the peak power requirement of an amplifier by reducing either the average output power or the peak-to-average ratio. In the latter case, intelligent signal clipping, whereby the baseband signals are limited in the digital domain, provides one possible solution. However, if such techniques are used, the effects on parameters such as the bit error rate (BER), vector error, and even the average power itself should be considered. System-level simulations are an important part of the total design process in this respect.

The frequency characteristic of the baseband transmitter filter, usually a *root-raised cosine* (RRC) characteristic with a *roll-off factor* α, also plays an important role in determining the peak-to-average ratio (Section 2.4.5). To reduce the level of adjacent channel power and hence interchannel interference, all baseband signals must be filtered; such filtering, however, causes a spreading of the signal in the time domain and gives rise to intersymbol interference (ISI).

In the case of a baseband signal consisting of rectangular pulses whose frequency content theoretically extends to infinity, filtering causes distortion of the signal. The time domain output of such a filter is no longer a perfect rectangular pulse since different frequency components are delayed by different amounts and a "ringing" effect in the waveform is observable. That is, the higher frequency components, which give the waveform its characteristic rectangular edges, are spread out in time. As the roll-off factor $\alpha \rightarrow 0$, the filter frequency characteristic more closely resembles a rectangular "brickwall filter," the group delay of the filter increases, and the amount of distortion (ringing) increases. Despite the increased distortion, low values of α are desirable since the level of adjacent channel power is reduced; to avoid ISI, the signals are sampled at a point when interference from preceding symbols is at a minimum. From a power amplifier perspective the opposite is true—higher values of α are better since the filter is less prone to ringing, which increases the signal peak-to-average ratio. The final choice of α is inevitably a compromise, and lower adjacent channel power is often favored over reduced peak-to-average ratio.

1.3.2.3 Calculating Signal Peak Power

The peak-to-average ratio, ΔP_s, of an input signal composed of N carriers, each having an average power P_i and peak-to-average ratio ψ_i can be calculated using (1.22), that is,

$$\Delta P_S = \frac{\left(\displaystyle\sum_{i=1}^{N} \sqrt{P_i \cdot \psi_i} \right)^2}{\displaystyle\sum_{i=1}^{N} P_i} \tag{1.22}$$

The peak power is calculated as the square of the sum of the individual carrier voltages, that is $(\sum V)^2$, while the average power is calculated as the sum of the square of the carrier voltages, that is $\sum (V^2)$.

For example, consider the following four scenarios:

1. Four GSM carriers each having an output power of 500 mW;
2. Four IS136 carriers each having an output power of 150 mW and a single 400-mW 1S95 carrier;
3. Twenty IS136 carriers each having an output power of 100 mW;
4. Two WCDMA carriers each having an output power of 15W.

GSM uses a constant envelope modulation scheme GMSK ($\psi = 1$) whereas the signal envelope for IS136 has a nonconstant envelope ($\psi = 2$). In IS95 and WCDMA, which use CDMA as the access technique, the peak-to-average ratio is a function of the number of codes; in these examples, the peak-to-average ratios of a single carrier are taken as 5 dB and 9 dB, respectively. The peak powers and peak-to-average ratios calculated using (1.22) are shown in Table 1.3.

The third example in particular demonstrates that the peak power rises rapidly as a function of the number of carriers. As shown in Chapter 2, the

Table 1.3
Peak Power and Peak-to-Average Ratio

Example	Number of carriers	Total average power	Peak power	Ratio ΔP_S
1	4 GSM	2W	8W	6 dB
2	4 IS136 1 IS95	1W	13W	11 dB
3	20 IS136	2W	80W	16 dB
4	2 WCDMA	30W	480W	12 dB

peak-to-average ratio in decibels increases on a logarithmic scale as the number of carriers increases (Figure 2.16). Alternatively, in a single-carrier CDMA system, the peak-to-average ratio could be shown as a function of the number of different codes rather than the number of carriers.

When the peak-to-average ratio of a signal is high due to the modulation scheme, the access technique, or the number of carriers, the peak power is also high. The true peak power is rarely reached in practice since all of the carriers (or codes) must be phase aligned. In practice, the random nature of the modulation data is such that this special case seldom happens. Furthermore, the duration of the peaks are inversely proportional to the peak-to-average ratio; that is, the peaks have a very short duration when the peak-to-average ratio is high. The practical peak power of a signal is therefore often taken as 10 times the average power, that is $\Delta P_S = 10$ dB. For example, in the previously described case of 20 IS136 carriers, the practical peak power is 20W compared to the true peak power of 80W. In the WCDMA example, the practical peak power would be 300W compared to a true peak power of 480W.

A multicarrier signal with a Rayleigh envelope distribution (Section 2.4.4) is, for example, characterized by a peak-to-average ratio, $\Delta P_S = 10$ dB, although the true peak power is theoretically infinite. A similar principle applies for single-carrier signals using complex modulation, for example, a CDMA signal using QPSK modulation. Instead of the theoretical maximum, the peak-to-average ratio could be quoted for a certain probability that the signal exceeds the peak value a certain percentage of the time. Such probability-based definitions lead to a more practical definition of the peak power since for high values of the peak-to-average ratio it is not economically viable to build amplifiers that have a peak power equal to the theoretical maximum.

1.3.2.4 Class of Operation

The class of operation of the main amplifier is essentially determined by the required average output power. For low-power applications (<10W), amplifiers can be used in Class A, which as discussed in Chapter 3, give good linearity but low efficiency. Low efficiency is less of a problem at low power than high power because the total heat dissipation is relatively low and the cooling requirements are more easily fulfilled. For low-power applications that require good linearity, Class A amplifiers are therefore used together with precorrection and/or feedforward techniques.

For higher power applications (>10W), it is important to minimize the power consumption; therefore, Class AB amplifiers, which have better efficiency but worse linearity than their Class A counterparts, are preferred. For example, amplifiers for cellular and PCS transmitters typically use Class AB amplifiers. Like Class A amplifiers, Class AB amplifiers can be used as the main amplifier in a feedforward system and/or combined with precorrection techniques. For efficiency reasons, at very high powers, such as required for satellite communications where the linearity requirements are less stringent than for terrestrial systems, amplifiers are often used in Class C.

1.3.2.5 Choice of Transistor

At RF, solid-state components, that is transistors, are used to generate RF power although some current research is focused on the suitability of traveling-wave tubes (TWTs) that are typically used for applications at higher microwave frequencies.

For example, the current choice of transistor technology for applications up to ≈3 GHz is between

- Silicon bipolars;
- Silicon MOSFETs;
- GaAs MESFETs.

GaAs transistors are typically operated in Class A, making them suitable for low-power applications; while Silicon bipolars and MOSFETs are used at higher powers in Class AB. The desired output power, linearity, efficiency, and frequency of operation are the primary considerations when choosing a transistor; as discussed in Chapter 3; however, there are many factors to consider.

1.3.2.6 Class AB Bipolar Transistor Line-Up

Figure 1.5 shows a typical transistor line-up for a 900-MHz/300-W amplifier using bipolar transistors; note that in practice, the peak power is slightly less than 300W due to the loss of the output combiner and dielectric (substrate) losses. The typical intermodulation performance and efficiency of bipolar transistors is discussed in Chapter 3 but is relatively constant at around −30 dBc for a Class AB amplifier.

Table 1.4 gives typical efficiency values as a function of output power and number of stages. For example, at an average power output of 30W, that is, 10 dB below the rated peak envelope power, the efficiency is approximately 15%, resulting in a DC power consumption of 200W.

Quadrature combining is nearly always used in the output stage of the amplifier to provide a good output match, such that reflections are minimized and parasitic signals incident at the output are attenuated. As will be discussed in Chapter 5, frequency ripples caused by impedance mismatch affects broadband signal cancellation.

The choice of substrate material is also an important factor, from both a performance and a production viewpoint. For example, it is important to minimize dielectric losses in the substrate material, while at the same time having material that is suitable for large-scale production. Woven or non-woven PTFE-based substrates have low losses but are not particularly suited

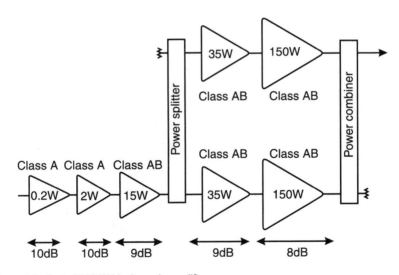

Figure 1.5 Typical 300-W bipolar main amplifier.

Table 1.4
Typical Efficiency of Class AB Bipolar Amplifiers

Power below PEP	Efficiency at 900 MHz (single stage)	Efficiency at 900 MHz (3 stages)
−3 dB	35	30
−4 dB	25	20
−10 dB	20	15

for large-scale production, whereas a material such as BTepoxy has the opposite characteristics. Note also that substrate losses increase with frequency.

1.3.2.7 Class AB MOSFET Transistor Line-Up

Figure 1.6 shows a transistor line-up for a 900-MHz/180-W amplifier using MOSFET transistors. As shown in the figure, MOSFETs typically have higher gain than their bipolar counterparts, resulting in fewer stages. The efficiency of a MOSFET amplifier is comparable to that of a bipolar amplifier (Table 1.5), although the intermodulation performance with MOSFETs is typically much better at around −40 dBc.

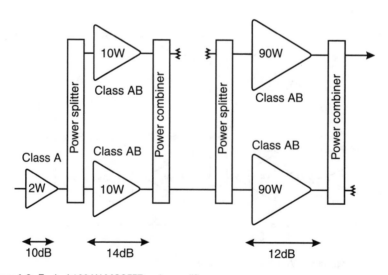

Figure 1.6 Typical 180-W MOSFET main amplifier.

Table 1.5
Typical Efficiency of Class AB MOSFET Amplifiers

Power below PEP	Efficiency at 900 MHz (single stage)	Efficiency at 900 MHz (3 stages)
–3 dB	35	30
–4 dB	25	21
–10 dB	16	13

1.3.2.8   Gain and Phase Flatness Requirements

The gain and phase flatness requirements of the main amplifier in a feed-forward system are typically those for 30-dB carrier suppression (Chapter 4):

- Gain flatness $< = \pm 0.2$ dB over the signal bandwidth;
- Phase flatness $< = \pm 1$ degree over the signal bandwidth.

The gain variation over signal level should also be minimized to reduce peak intermodulation due to loop unbalance, that is $\Delta G < = 2$ dB over the signal dynamic range. Note that the gain variation is typically much lower with MOSFETs than with bipolars when the intermodulation performance has been optimized.

1.3.3   Error Amplifier

As explained in Chapter 5, the requirements on the error amplifier in a feed-forward system differ in several respects from those for the main amplifier. Briefly summarizing, compared to the main amplifier, the error amplifier is required to have better linearity, more broadband performance, and lower average and peak power capability. For practical reasons the peak power requirements of an amplifier should be kept as low as possible; therefore, it is desirable to have the following general conditions fulfilled:

- Good intermodulation performance from the main amplifier, that is, minimum intermodulation power in the error amplifier path;
- Low main amplifier average output power;
- Good loop suppression, that is, minimum residual carrier power;
- Low signal peak-to-average ratio.

The peak power of the main amplifier is determined by the desired average output power times the signal peak-to-average ratio. The average output power is in turn related to the system configuration and the peak-to-average ratio is determined by the statistics of the input signal, that is, the modulation format and number of carriers. Intermodulation performance of the main amplifier is related to the choice of transistor, class of operation, average output power, and frequency of operation. Residual carrier power is a function of the broadband loop cancellation—that is, the accuracy of the gain, phase, and delay matching; the amount of phase and amplitude ripple across the band; and the carrier bandwidth. Any change in these parameters results in modified average and peak power requirements on the error amplifier.

1.3.3.1 Linearity, Efficiency, and Class of Operation

Efficiency and linearity characteristics of Class A and AB amplifiers are described in Chapter 3, and the linearity requirements placed on the error amplifier are such that Class A operation is nearly always chosen. This is because any intermodulation created by the error amplifier arrives at the feedforward output and contributes directly to the output distortion. There is no possibility of cancellation unless a second feedforward loop is used—even then, the same argument applies to the error amplifier in the outermost loop. The target for error amplifier linearity depends on the required output linearity and the value of the output coupler can be as high as −60 dBc. The choice of Class A operation means that bipolar and GaAs transistors are those most commonly used in error amplifiers.

1.3.3.2 Bipolar Error Amplifier

Figure 1.7 shows a typical, high-gain, bipolar error amplifier. Since the transistors are biased in Class A, the power-handling capability and intermodulation performance are typically described by the 1-dB compression (P_{1E}) and third-order intercept point ($IP3$); see Chapter 3.

Typical performance figures are:

- Efficiency at P_{1E}: $\eta_E = 20\%$;
- Third-order intercept: $IP3 = P_{1E} + 10$ dB.

Note that shielding considerations are particularly important in high-gain amplifiers; the individual stages may be physically very close to each

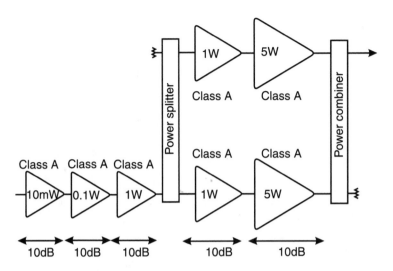

Figure 1.7 Typical 10-W bipolar error amplifier.

other, thus increasing the risk of unwanted feedback and instability. Furthermore, compared to the main amplifier and other parts of the feedforward circuit, the signal levels in the error amplifier are very small and thus the isolation must be correspondingly high. For example, if a small portion of a pilot signal inadvertently "leaks" into the error amplifier, the loop control network can incorrectly align the loop and consequently degrade the feedforward output linearity.

1.3.3.3   GaAs Error Amplifier

Like MOSFETs, GaAs MESFETs typically have higher gains than bipolars, resulting in amplifiers with fewer stages (Figure 1.8). Typically, GaAs MESFETs have a higher efficiency than bipolars and, furthermore, the third-order intercept point at low levels is higher:

- Efficiency at P_{1E}: $\eta_E = 30\%$;
- Third-order intercept: $IP3 = P_{1E} + 15$ dB.

The overall efficiency of a feedforward amplifier is related to the efficiency of the error amplifier (Chapter 5); hence, GaAs is often preferred to bipolar.

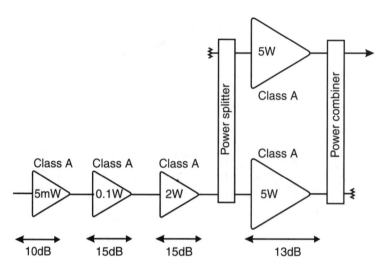

Figure 1.8 Typical 10-W GaAs error amplifier.

1.3.3.4 Gain and Phase Flatness Requirements

The gain and phase flatness requirements on the error amplifier depend upon the required loop suppression, which in turn is a function of the intermodulation performance of the main amplifier and the desired feedforward output linearity. For example, if the main amplifier has an intermodulation level of −30 dBc and the required output distortion level is −65 dBc, the loop suppression should be 35 dB. Gain and phase flatness as a function of loop suppression has already been discussed; however, in regard to the error amplifier it is important to realize that the cancellation bandwidth is much wider than in carrier cancellation loops. For example, the distortion cancellation bandwidth is typically three times the signal bandwidth.

1.3.4 Directional Couplers

As previously mentioned, there are several occasions when power needs to be combined or divided in a feedforward amplifier. For example, power is divided at the feedforward input and at the output of the main amplifier and power is combined (in antiphase) for carrier and distortion cancellation (Chapter 4). Note that the *same* device can be used as a power splitter or combiner; the functionality is determined only by how the device is connected.

Although power splitters and combiners can be implemented using resistive dividers, directional couplers are nearly always used in practice because
they can be designed to give good isolation, impedance match, and insertion
loss and have a wide operating bandwidth. Resistive dividers are inherently
lossy and are very sensitive to mismatch errors; even a small mismatch can
give rise to an appreciable voltage standing wave ratio.

1.3.4.1 Feedforward Input Coupler

The power splitter at the feedforward input separates the input signal into
two paths: one going to the main amplifier, the other to a delay element. The
reference path is normally chosen to have the lowest loss since this gives a
better feedforward noise performance (Chapter 5). For example, if a 10-dB
coupler is used as the input power splitter, then the insertion loss in the reference path is theoretically 0.5 dB, that is, only a 0.5-dB contribution to the
overall noise figure. The higher loss in the main amplifier path (10 dB in this
case) is compensated for by an appropriate increase in gain before the main
amplifier.

1.3.4.2 Main Amplifier Output Coupler

The high-power output of the main amplifier is sampled to provide a cancellation signal for the first feedforward loop. For reasons previously discussed,
the objectives are to minimize insertion loss and have high directivity (high
isolation). For example, a 20-dB coupler has a theoretical insertion loss of
only 0.1 dB and a directivity of >15 dB is achieved if the isolation is >35 dB.

1.3.4.3 Loop 1 (Carrier-Cancellation Coupler)

The loop 1 cancellation coupler is an example of a power combiner; the
power is actually added in antiphase and therefore the functionality is that of
a subtractor. Since the carrier signals can be suppressed in excess of 30 dB
(comparable to the intermodulation level), the power levels at the coupler
output are very low. Good directivity is important to minimize amplitude
frequency ripple and to ensure a high input return loss (Chapter 5); low insertion loss is desirable for good, overall noise performance.

1.3.4.4 Loop 2 (Distortion-Cancellation Coupler)

As in loop 1, the cancellation coupler in loop 2 is also an example of a power
combiner and has the functionality of a subtractor; the difference is that
distortion components are canceled rather than the carriers. In loop 2 there is
a trade-off between insertion loss and coupling factor since low insertion loss
implies a high coupling factor and hence a more powerful error amplifier.

Conversely, a low coupling factor allows the error amplifier to be smaller but requires a higher output power from the main amplifier. The optimum coupling value is derived in Chapter 5, and typically the coupling coefficient is equal to the ratio (dB) of the main and error amplifier power-handling capabilities.

1.3.5 Delay Lines

The purpose of the delay elements in a feedforward amplifier is to make signal cancellation independent of frequency. The most important performance parameters of delay lines are insertion loss and frequency response, that is, constant gain and linear phase (constant delay), although other factors—such as behavior over temperature, physical size, cost, and suitability for production—are also important. Electrical delay, linear phase, and broadband signal cancellation are discussed in more detail in Chapters 4 and 5; the emphasis here is placed on insertion loss and the practical realization of delay elements.

1.3.5.1 Delay Line Loss

The subject of transmission lines and the concept of distributed parameters and characteristic impedance has already been introduced (Section 1.1.5). Another important parameter of a transmission line is the transmission line propagation constant γ (a complex quantity having real and imaginary terms), that is

$$\gamma = \alpha + j \cdot \beta \tag{1.23}$$

The real part a is called the *attenuation constant* (units dB/m or Nepers/m) and is equal to zero for a *lossless* line. The imaginary term β is called the *phase constant* (units radians/meter) and describes the phase shift per unit length. If β is a linear function of frequency, the phase is linear and hence the delay is constant.

Equation (1.23) can be rewritten in terms of the transmission line constants R, L, G, and C, which represent the series resistance, inductance, shunt capacitance, and conductance per unit meter, respectively, that is

$$\gamma = \sqrt{(R + j \cdot \omega \cdot L)(G + j \cdot \omega \cdot C)} \tag{1.24}$$

The series resistance R is a measure of the conductor loss, while the shunt conductance G is a measure of the dielectric loss. As previously mentioned, dielectric losses are directly proportional to frequency and are normally described by the loss tangent, tan δ. One method of reducing dielectric loss in, for example, a coaxial cable, is to reduce the amount of dielectric material between the inner and outer conductors. For example, a helical spiral can be used to support the inner conductor rather than a solid dielectric. Loss in a cable can also be reduced by increasing the diameter of the cable; large-diameter cables are difficult to handle for mechanical reasons and are very expensive.

1.3.5.2 Skin Depth

In a good conductor such as copper, sinusoidal waves are heavily attenuated and therefore do not penetrate very far into the conductor; waves tend to flow on the surface in a phenomenon referred to as the *skin effect.* In this special case, the attenuation constant and phase constant are numerically equal, that is,

$$\alpha = \beta = \sqrt{\pi \cdot f \cdot \mu \cdot \sigma} \qquad (1.25)$$

Conductor losses are thus proportional to $\sqrt{f}$ and the distance a wave travels before its amplitude is reduced by a factor e^{-1} or 0.368 is known as the skin depth δ, that is,

$$\delta = \frac{1}{\alpha} = \frac{1}{\sqrt{\pi \cdot f \cdot \mu \cdot \sigma}} \qquad (1.26)$$

For example, the skin depth of copper at 100 MHz is 6.6 μm while at 2 GHz, the skin depth is only 1.5 μm. In high-frequency applications, the condition (clean and tarnished, e.g.) of the surface of conductors and connectors is therefore very important since this is where the current flows. Copper oxide, which inevitably forms on exposure to the air, is relatively lossy; hence, silver or gold plating is sometimes used for connectors and signal tracks.

1.3.5.3 Delay Elements

There are a number of different methods for introducing delay into a network and each one has advantages and disadvantages in terms of loss, cost, and ease of manufacture. For example, Table 1.6 shows the loss of several different delay structures, each having the same delay, that is, 10 ns at 1 GHz (remember that losses increase with frequency and delay).

Table 1.6
Loss in Delay Line Structures

Structure	Approx. Loss@1 GHz/10 ns
Bandpass cavity filter	0.2 dB
Suspended line	0.5 dB
Semirigid coaxial cable	
250 mils	0.5 dB
140 mils	0.8 dB
Stripline	0.9 dB

The cavity filter has the lowest loss but since this has a bandpass rather than a lowpass characteristic, the delay is only constant well within the passband. Semirigid cables, which have a lowpass and therefore constant delay characteristic, are particularly suited for use as delay elements in feedforward amplifiers although stripline structures offer comparable loss. The final choice of delay line varies from one implementation to another; however, coaxial cables are probably the most common form of delay lines in feedforward amplifiers.

2

Linearity and Signal Description

2.1 Linear Amplifier Input/Output Characteristics

Figure 2.1a shows an amplifier represented as a two-port network having an input voltage $V_{in}(t)$, output voltage $V_{out}(t)$, and transfer function $T(\omega)$. For a perfectly linear amplifier, the output voltage is simply a constant times the input voltage, that is,

$$V_{out}(t) = G \cdot V_{in}(t) \tag{2.1}$$

Regardless of signal level, all signals are increased in magnitude by the same factor G and there is a fixed phase shift (equal to the time delay) between input and output for a signal at a given frequency.

In terms of the frequency response $T(\omega)$, an ideal amplifier has constant characteristics over the *bandwidth* of the input signal—that is, constant gain, linear phase, and hence, constant delay. An ideal amplifier is also memoryless; that is, the response of the amplifier at any point in time is determined solely by the value of the input signal at that moment and not by any previous values.

In practice, however, the devices used in amplifiers, such as transistors, have nonlinearities that make the output voltage a function of higher order terms of the input voltage; the input/output characteristic is then said to be *nonlinear*. Such nonlinear amplifiers also typically have frequency-dependent gain, nonlinear phase, and memory.

43

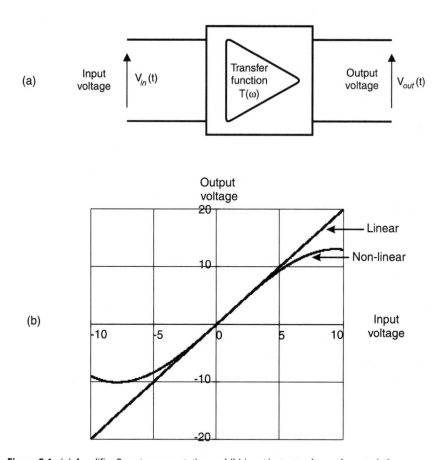

Figure 2.1 (a) Amplifier 2-port representation and (b) input/output voltage characteristic.

2.1.1　Series Representation of a Nonlinear Amplifier

A nonlinear output voltage can be expressed mathematically as a Taylor series such that

$$V_{\text{out}}(t) = G_1 \cdot V_{\text{in}}(t) + G_2 \cdot V_{\text{in}}^2(t) + G_3 \cdot V_{\text{in}}^3(t) + \cdots G_n \cdot V_{\text{in}}^n(t) \quad (2.2)$$

The amplifier constants $G_{1..n}$ determine the exact shape of the input/output characteristic; for example, Figure 2.1(b) shows a voltage transfer function having the form given by (2.2) (for comparison, the linear response (2.1) is also shown).

As Figure 2.1(b) illustrates, at high signal levels the output voltage compresses for both positive and negative values. This type of compression (or signal clipping) is due to the third-order term G_3 while the second-order term G_2 tends to cause overshoot at one end (gain expansion) and clipping (gain compression) at the other. In practice, both terms are present to a greater or lesser extent, resulting in distortion of the output signal regardless of the input signal level.

2.1.2 AM-AM and AM-PM Characteristics

Voltages are vector quantities having both amplitude and phase; therefore, an alternative way of looking at the input/output characteristic is to treat amplitude and phase separately. This method is similar to that used for frequency transfer functions, which also have a complex amplitude and phase response—the difference there is that the amplitude and phase responses are functions of frequency and not input level.

For example, Figures 2.2(a,b) show the amplifier input/output characteristic in terms of the amplitude and phase response for the same case as in Figure 2.1. The amplitude response (Figure 2.2(a)) is referred to as the *AM-AM characteristic* and the phase response (Figure 2.2(b)) is called the *AM-PM characteristic*. The distortion introduced by a nonlinear amplifier is frequently explained in terms of AM-AM and AM-PM characteristics and is strongly

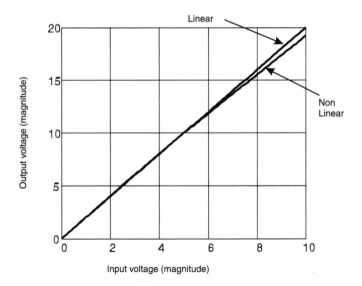

Figure 2.2 (a) Amplifier AM-AM characteristic.

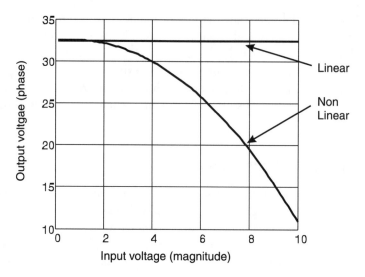

Figure 2.2 (continued) (b) AM-PM characteristic.

dependent upon the class of operation in which the amplifier is used (see Chapter 3 for details).

2.1.3 Single-Carrier Output and Harmonic Distortion

Consider a single unmodulated CW carrier as the input signal; the input voltage (peak amplitude a, frequency f_1, and arbitrary phase offset ϕ) then has the form

$$V_{in}(t) = a \cdot \cos(2 \cdot \pi \cdot f_1 \cdot t + \phi) \tag{2.3}$$

The linear output voltage is calculated from (2.1) and the nonlinear output voltage from (2.2). Figures 2.3(a) and 2.4(a) show these signals in the time domain; as expected, the nonlinear transfer function causes signal clipping (compression) of the output voltage.

The frequency-domain response, which is obtained by taking the *Fourier transform* of the time-domain waveforms, is usually presented in the form of a power and phase spectrum, that is,

$$\text{Power}(f) = \xrightarrow{\left(\left\|\text{fft}(v_{out})\right\|\right)^2}$$

$$\text{Phase}(f) = \xrightarrow{\arg(\text{fft}(v_{out}))} \tag{2.4}$$

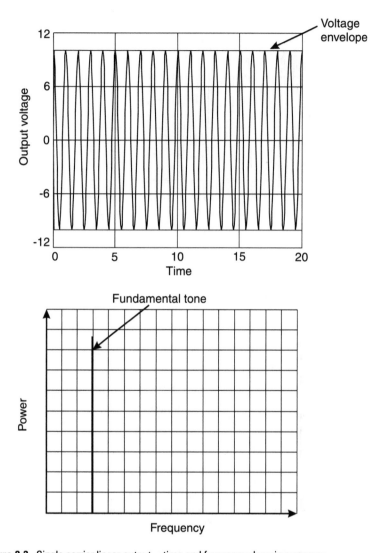

Figure 2.3 Single carrier linear output—time and frequency domain response.

Figures 2.3(b) and 2.4(b) show the power spectra for linear and nonlinear outputs, respectively. In the linear case only the amplified frequency at f_1 is present, while in the nonlinear example there are additional frequency terms, namely:

- A DC component;
- The fundamental tone f_1—amplitude compressed;

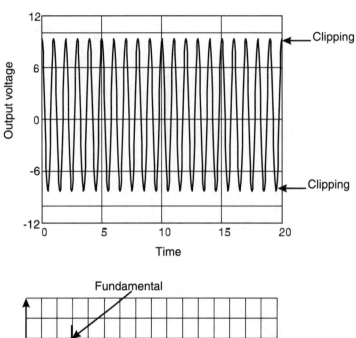

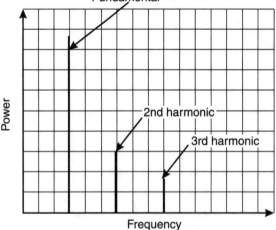

Figure 2.4 Single carrier nonlinear output—time and frequency domain response.

- The second harmonic at $2f_1$;
- The third harmonic at $3f_1$.

The response shown in Figure 2.4(b) is an example of *harmonic distortion* and occurs even with a relatively simple signal such as a single unmodulated carrier. Alternatively as Table 2.1 shows, the same result can be

Table 2.1
Harmonic Distortion

DC term	$\dfrac{G_2 \cdot a^2}{2}$
Fundamental (& compression term)	$G_1\left(1 + \dfrac{3 \cdot G_3 \cdot a^2}{4 \cdot G_1}\right) \cdot a \cdot \cos\theta$
Second harmonic	$\dfrac{G_2 \cdot a^2}{2} \cdot \cos 2\theta$
Third harmonic	$\dfrac{G_3 \cdot a^3}{4} \cdot \cos 3\theta$

presented in a different form by evaluating in the time domain with $\theta = 2\pi f_1 t$ and collecting terms of similar order.

Note that for a 1-dB increase in input signal level, the second-order harmonic term goes up by 2 dB (proportional to a^2) and the third-order harmonic by 3 dB (proportional to a^3).

2.1.4 Two-Tone Test—Harmonic and Intermodulation Distortion

Now consider two tones of equal amplitude "a" having frequency f_1 and f_2, respectively, that is,

$$V_{in}(t) = a \cdot \cos(2 \cdot \pi \cdot f_1 \cdot t) + a \cdot \cos(2 \cdot \pi \cdot f_2 \cdot t) \qquad (2.5)$$

Figure 2.5(a) shows the time-domain waveform for linear output and it is evident that the signal envelope is no longer constant as in the single-carrier case but varies between maximum and minimum values. This particular type of nonconstant envelope behavior makes the two-tone test a very useful signal for test and measurement purposes since the amplifier is driven through the whole range of its transfer characteristic (from zero to the signal envelope maximum). There is also an important practical advantage associated with a two-tone test, the ease of signal generation.

For a nonlinear output (Figure 2.6(a)), the signal no longer follows the true envelope shape and there is asymmetrical signal clipping resulting in distortion. The Fourier transform representation of this distorted time-domain waveform is shown in Figure 2.6(b) and, in addition to harmonic

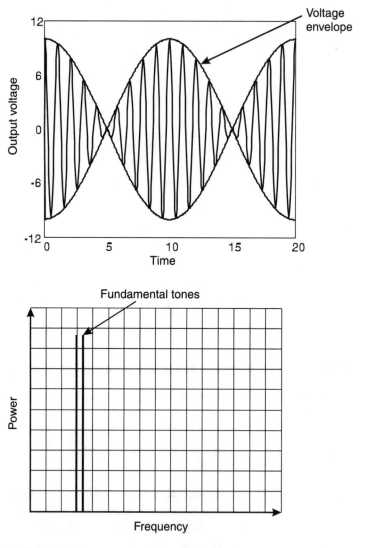

Figure 2.5 Two-carrier linear output—time and frequency domain response.

distortion, other frequency components or *intermodulation (IM) products* are also present.

Thus, for two unmodulated tones the frequency spectra consists of:

- A DC term;
- Fundamental tones f_1 and f_2—compressed;

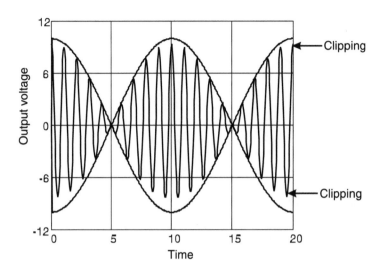

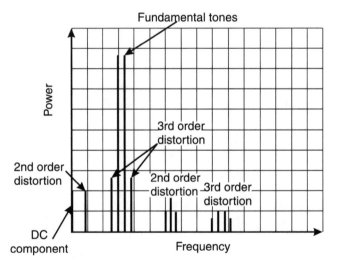

Figure 2.6 Two carrier nonlinear output—time and frequency domain response.

- Harmonics;
- Intermodulation products.

Alternatively, Table 2.2 shows the results of evaluating in the time domain with a two-tone signal as the input and collecting terms of similar order.

Table 2.2
Two-Tone Test Frequency Components

DC term	$G_2 \cdot a^2$
Fundamental	$G_1 \cdot a \cdot \left(1 + \dfrac{9 \cdot G_3 \cdot a^2}{4 \cdot G_1}\right) \cdot (\cos \theta_1 + \cos \theta_2)$
Second order	$\dfrac{G_2 \cdot a^2}{2} \cdot (\cos 2\theta_1 + \cos 2\theta_2) \cdots$ $+ G_2 \cdot a^2 \cdot \big(\cos(\theta_1 + \theta_2) + \cos(\theta_1 - \theta_2)\big)$
Third order	$\dfrac{G_3 \cdot a^3}{4} \cdot (\cos 3\theta_1 + \cos 3\theta_2)$ $+ \dfrac{3 \cdot G_3 \cdot a^3}{4} \big[\cos(2 \cdot \theta_1 + \theta_2) + \cos(2\theta_1 - \theta_2)\big] \cdots$ $+ \dfrac{3 \cdot G_3 \cdot a^3}{4} \cdot \big(\cos(2\theta_2 + \theta_1) + \cos(2 \cdot \theta_2 - \theta_1)\big)$

As before, if the input signal level is increased by 1 dB, the second-order terms increase by 2 dB (proportional to a^2) and the third-order terms increase by 3 dB (proportional to a^3).

2.1.5 Third-Order Intercept Point (IP3)

In order to characterize the third-order distortion of an amplifier, the terms $\cos(2\theta_1 - \theta_2)$ and $\cos(2\theta_2 - \theta_1)$ are often used. Since they are proportional to a^3, these intermodulation products increase by 3 dB when the fundamental goes up by 1 dB. The third-order intercept point is then defined as the theoretical level at which the intermodulation products are equal to the fundamental tone (Figure 2.7).

2.1.6 Distortion of Multicarrier Signals

For a multicarrier signal composed of evenly spaced tones (spacing Δf), the intermodulation products also fall on a Δf grid. IM products thus appear within the same band as the carriers themselves, and hence any thoughts

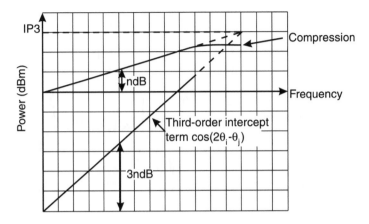

Figure 2.7 Third-order intercept point (IP3).

about using a filter to remove unwanted intermodulation products must now clearly be abandoned.

As the number of tones is increased, the number of third-order beats (IM products due to third-order distortion) also increases and, theoretically, the highest intermodulation level occurs in the center of the band. In order to measure this "worst case," a gap is often left in the middle of the carriers. For example, Figure 2.8 shows two groups of four tones separated by a gap of $2\Delta f$ and the intermodulation product in the center is clearly visible (see Figure 2.9). Note that if the number of tones is increased but the total average power is kept constant, the intermodulation performance is degraded.

2.1.7 Intermodulation and Spectral Regrowth

The nonlinearities produced by discrete signals such as unmodulated carriers are known as intermodulation products and appear at discrete frequencies. For more complex (modulated) signals, however, the nonlinearities appear over a continuous band of frequencies and are often referred to as *spectral regrowth*. For example, the level of adjacent channel power (a measure of the spectral regrowth) is often used as a measure of linearity for complex modulated signals as opposed to the level of discrete intermodulation products for a simple two-tone test.

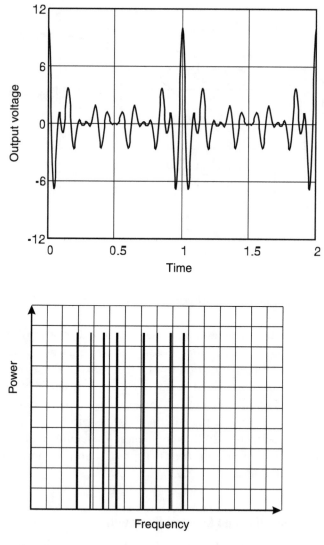

Figure 2.8 Eight-carrier linear output—time- and frequency-domain response.

2.2 Modulation Formats

Modulation techniques can be broadly separated into three groups and the resulting signal waveforms into two groups (constant and nonconstant envelope signals), that is,

1. Constant envelope modulation schemes, which modulate *one parameter* of a carrier wave and result in a signal with a *constant envelope*.

2. Nonconstant envelope modulation schemes, which modulate one parameter of a carrier wave and result in a signal that has a nonconstant envelope.

3. Linear modulation schemes, which modulate amplitude and phase together, resulting in a signal having a nonconstant envelope.

For example, analog frequency modulation schemes belong to the first group and amplitude modulation to the second. Digital phase and frequency modulation schemes such as phase shift keying (PSK), minimum shift keying (e.g., GMSK), and frequency shift keying (FSK) belong to the first group, and digital linear modulation schemes such as Quadrature Amplitude Modulation (QAM) belong to the third group.

Signals modulated with GMSK have a constant envelope (no information is transmitted in the amplitude of the signal); however, in the particular case of GSM, which is a TDMA system, signals have a nonconstant envelope due to power regulation between timeslots. PSK signals (e.g., QPSK) also have a nonconstant envelope even though no information is transmitted in the amplitude of the signal; amplitude variations are introduced during

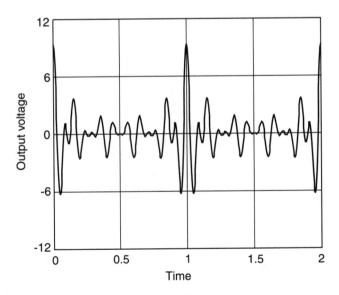

Figure 2.9 (a) Eight-carrier nonlinear output—time-domain response.

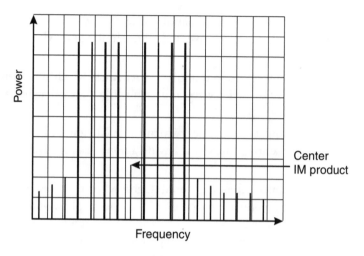

Figure 2.9 (continued) (b) Frequency domain response.

transitions between points in the constellation diagram (Figure 2.10). In order to minimize signal envelope variations the modulation scheme can be modified (e.g., QPSK to $\pi/4$-QPSK), but amplitude variations cannot be completely removed.

The third group contains those modulation techniques that use both amplitude and phase to convey information. An example of such a linear modulation scheme is 16-QAM and the constellation diagram is shown in

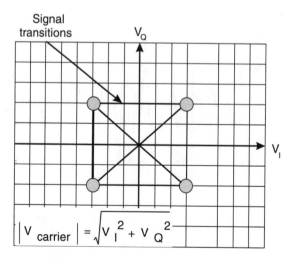

Figure 2.10 QPSK constellation diagram.

Figure 2.11. Linear modulation schemes are inherently more spectrally efficient than single-parameter modulation schemes, but since the amplitude is being varied, the envelope of such a signal is not constant.

2.2.1 Spectral Efficiency

Spectral efficiency is important because the electromagnetic spectrum is a finite natural resource and the available bandwidth for radio systems is limited. The spectral efficiency of a radio system can be defined in terms of its throughput per unit frequency, that is, the number of bits/second per Hertz, and obviously high spectral efficiency is achieved when high data rates can be supported in a narrow bandwidth. It is thus important to consider not just the bandwidth of a particular signal but also its potential throughput capability. For example, one radio channel may have a bandwidth of 25 kHz, but this does not necessarily mean that it is more spectrally efficient than a 200-kHz or even 1.25-MHz channel using a different modulation technique—a more detailed analysis is required before such a statement can be made.

2.3 Signal Envelopes

Figure 2.12(a) shows an unmodulated CW signal where the voltage envelope is constant as a function of time but could also represent a true constant enve-

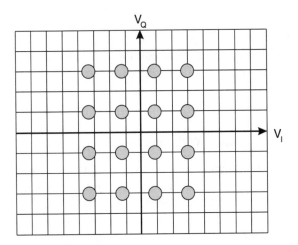

Figure 2.11 Linear modulation constellation diagram (16-QAM).

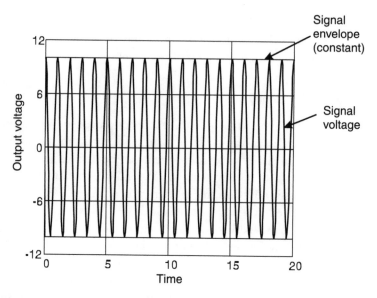

Figure 2.12 (a) Constant envelope signal.

lope modulated signal; for a comparison, a signal with a nonconstant envelope (two CW tones) is shown in Figure 2.12(b).

2.3.1 Constant Envelope Signals

In terms of the linearity requirements on the power amplifier, the amplification of a single constant-envelope signal (Figure 2.12(a)) is relatively straightforward. First, the envelope peak power is equal to the envelope average or
mean power; and second, the signal envelope is at a fixed point in the amplifier's transfer function, thus eliminating the effects of AM-AM and AM-PM
distortion.

2.3.2 Multicarrier Signals With Constant Envelope Modulation

Regardless of the modulation scheme, as soon as there is more than one carrier, the signal envelope varies and this has important consequences in terms
of linearity and power amplifier design. For example, consider the case of N
unmodulated carriers with uniform frequency spacing $\Delta\omega$, each having unity
amplitude and zero phase-offset. The multicarrier signal, which is the vector
summation of the individual signals, is then written as

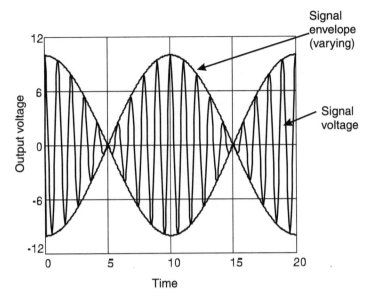

Figure 2.12 (continued) (b) Non-constant envelope signal.

$$V = \sum_{n=0}^{N-1} 1 \cdot \cos(\omega \cdot t + n \cdot \Delta\omega \cdot t) \qquad (2.6)$$

or, alternatively,

$$V = \cos\left[\omega \cdot t + \frac{(N-1) \cdot \Delta\omega \cdot t}{2}\right] \cdot \frac{\sin\left(\dfrac{N \cdot \Delta\omega \cdot t}{2}\right)}{\sin\left(\dfrac{\Delta\omega \cdot t}{2}\right)} \qquad (2.7)$$

The first term of (2.7) is an RF signal having a frequency equal to the midband or center frequency of the carriers; the second term describes the amplitude modulation of the RF signal. From (2.7), the envelope of the multitone signal is therefore

$$V_{env} = \left|\frac{\sin\left(\dfrac{N \cdot \Delta\omega \cdot t}{2}\right)}{\sin\left(\dfrac{\Delta\omega \cdot t}{2}\right)}\right| \qquad (2.8)$$

An N-carrier signal can be thought of as the summation of N rotating vectors (phasors) having a frequency difference equal to the carrier spacing Δf, and peaks occur when the rotating phasors have the same phase. The peak voltage of the multicarrier signal is then equal to N times the peak voltage of a single carrier and the carriers are said to be *phase aligned*. That is, the normalized peak voltage P_{peak} is given by

$$P_{peak} = \left(\sum_{i=0}^{N-1} V_{peak_i} \right)^2 \qquad (2.9)$$

Figure 2.13 shows the envelope waveform for a 16-carrier signal and peaks occur when the denominator of (2.8) is equal to zero, that is,

$$\sin\left(\frac{\Delta\omega \cdot t}{2} \right) = 0 \qquad (2.10)$$

Solving for $t = T_{env}$ gives $T_{env} = 1 / \Delta f$, that is, the peaks occur at regular intervals equal to the reciprocal of the carrier spacing Δf. The width of the peaks can be calculated by seeing when the first zero occurs, that is, when

$$\frac{N \cdot \Delta\omega \cdot t}{2} = \pi \qquad (2.11)$$

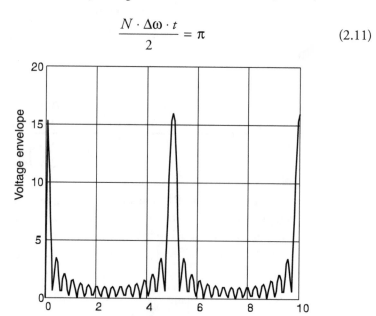

Figure 2.13 Voltage envelope—16 carriers phase aligned.

Solving for $t = T_{zero}$ gives $T_{zero} = 1 / N\Delta f$ and the peak width is thus proportional to $1 / N\Delta f$. For example, if $f_1 = 900$ MHz and $f_{16} = 915$ MHz ($\Delta f = 1$ MHz), the composite 16-tone unmodulated signal has peaks of width 0.0625 ms evenly spaced at 1-ms intervals.

Although the peaks have a very short duration, an amplifier must be designed for the peak-power level ($\propto N^2$) to avoid distortion even though the amplitude levels are low for the majority of time. As shown in Chapter 3, the efficiency of an amplifier is, in general, low at low amplitude levels; hence, a signal with a high peak-to-average ratio will result in an overall low efficiency.

If the carriers are not phase-aligned but rather have random phases, as is the case in practice, then the voltage vectors for individual carriers are unlikely to all have a maximum at the same time and the peaks are reduced in amplitude. This lowering of the peak amplitude is demonstrated clearly in Figure 2.14, which shows the same 16-carrier example for the cases of random and peaked carrier phases.

2.3.3 Nonconstant Envelope Signals

For nonconstant envelope signals, (2.6) is modified to become

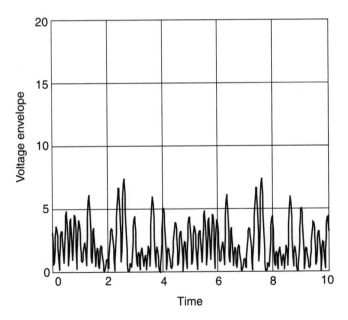

Figure 2.14 Voltage envelope—16 carriers with random phase.

$$V = \sum_{n=0}^{N-1} a_n(t) \cdot \cos\left(\omega \cdot t + n \cdot \Delta\omega \cdot t + \phi_n(t)\right) \qquad (2.12)$$

Note that this description applies for single- and multicarrier signals. For example, as previously mentioned, the amplitude of a single QPSK modulated carrier varies even though only the phase (ϕ_n) is used to convey information. With linear modulation schemes, both the amplitude and phase, a_n and ϕ_n, contain information as the envelope amplitude varies in response to the transmitted data. For example, Figure 2.15 shows a linearly modulated 16-carrier signal and the similarities with Figure 2.14 (random phase, constant envelope modulation) are clear: the peaks occur "randomly" and are generally reduced in amplitude.

Transmitted data is normally considered to be a random variable and hence, carrier amplitude and phase can also be considered to be random variables. A precise description of the peaks for a nonconstant envelope signal can therefore only be found using probability theory (e.g., distribution and density functions), and this leads to the definition of an important parameter called the *peak-to-mean* or *peak-to-average ratio*.

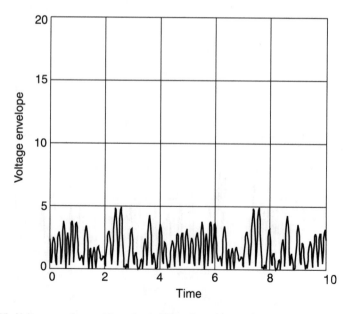

Figure 2.15 Voltage envelope—16 carriers with random phase and amplitude.

2.4 Peak-to-Average Ratio

The ratio between the peak and the average power of a signal is called the peak-to-average or peak-to-mean ratio, ΔP_s. All signals (single- or multicarrier, modulated or unmodulated) have a peak-to-average ratio; for example, a single unmodulated carrier has a peak-to-average ratio of unity ($\Delta P_s = 0$ dB) while multicarrier or complex modulated signals can have peak-to-average ratios in excess of 10 dB.

Peak-to-average ratios are not restricted to signals; however, amplifiers can also have a peak-to-average ratio. In the case of an amplifier the peak power is normally given as the rated peak-envelope-power (PEP) or the 1-dB compression point (see Chapter 3). As an example, for an amplifier with an average output power P_M and 1-dB compression point P_{1M}, the peak-to-average ratio ΔP_M equals P_{1M} / P_M.

Peak power and hence peak-to-average ratio, which is used to calculate peak power, are very important for linearity reasons. If the peaks of a signal are greater than the peak power of the amplifier, then signal clipping occurs, causing distortion of the output signal in the form of intermodulation distortion and spectral regrowth. Amplifiers are thus normally designed such that their peak-power capability is at least equal to the signal peak power. As will be explained, definitions of peak power are often quoted in combination with probability density functions.

2.4.1 Signal and Envelope Peak-to-Mean Ratio

The peak-to-average ratio of a signal can be described in terms of either the *signal* peak-to-average or the *envelope* peak-to-average.

Referring back to Figure 2.12, the signal peak power and the envelope peak power are equal (the peak voltage is the same in both cases). Furthermore, from Chapter 1, the peak power is calculated as the *square of the voltage sum*, that is,

$$P_{peak} = \frac{\left(\sum_N V_{max} \right)^2}{R} = \frac{N^2 \cdot V_{max}^2}{R} \tag{2.13}$$

As previously indicated, a multicarrier signal can be thought of as the summation of independent random variables (voltages) each with mean or average voltage V_{rms} and power V_{rms}^2. From probability theory, the total average power P_{signal_av} of the multicarrier signal is then given by

$$P_{signal\_av} = \sum_N \frac{V_{rms}^2}{R} \tag{2.14}$$

That is, the average power is calculated as the *sum of the voltage squares*. Substituting for V_{rms} (assuming sinusoidal excitation) gives

$$P_{signal\_av} = \frac{1}{R} \cdot \sum_N \left(\frac{V_{max}}{\sqrt{2}}\right)^2 = \frac{N \cdot V_{max}^2}{2 \cdot R} \tag{2.15}$$

The *signal* peak-to-average ratio is therefore equal to

$$\frac{P_{peak}}{P_{signal\_av}} = 2 \cdot N \tag{2.16}$$

The average power of the envelope can be obtained by evaluating (2.8) over one cycle, that is,

$$P_{env\_av} = \frac{1}{T} \cdot \int_0^T \frac{V_{env}^2(t)}{R} dt \tag{2.17}$$

Substituting for V_{env} in (2.17) and evaluating gives

$$P_{env\_av} = \frac{1}{R \cdot T} \cdot \int_0^T \left[\left|\frac{\sin\left(\frac{N \cdot \Delta\omega \cdot t}{2}\right)}{\sin\left(\frac{\Delta\omega \cdot t}{2}\right)}\right| \cdot V_{max}\right]^2 dt = \frac{N \cdot V_{max}^2}{R} \tag{2.18}$$

The envelope peak-to-average ratio is thus

$$\frac{P_{peak}}{P_{env\_av}} = N \tag{2.19}$$

Note that the envelope ratio, which is 3 dB lower than the signal peak-to-average ratio is that typically used when referring to multicarrier signals.

Figure 2.16 shows the envelope peak-to-average ratio (dB) as a function of the number of unmodulated carriers N. For example, if $N = 32$ the peak-to-average ratio is 15 dB (assuming all carriers are phase aligned with maximum amplitude).

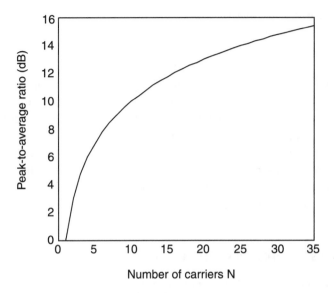

Figure 2.16 Peak-to-average ratio and number of carriers.

2.4.2 Amplitude Density Functions

A random or *stochastic* variable can be formally defined as

> The random variable X is a function associated with an experiment whose values are real numbers and their occurrence in trials depends on "chance."

A continuous random variable, such as a voltage V with *amplitude* (probability) density function $f(v)$, has the property that

$$\int_{-\infty}^{\infty} f(v)\, dv = 1 \qquad (2.20)$$

That is, the area under the curve of $f(v)$ is equal to unity (one of the values must occur). Note also that $f(v)$ is a probability *density* function and must be integrated between two limits to get the absolute probability, that is, the probability that V lies between value v_1 and v_2 is given by

$$P(v_1 < V < v_2) = F(v_2) - F(v_1) = \int_{v_1}^{v_2} f(v)\, dv \qquad (2.21)$$

where $F(v)$ is the cumulative distribution function or simply the distribution function of the random variable V.

Neither the amplitude density function $f(v)$ nor the distribution function $F(v)$ completely describes a signal (its spectral density is also required), but they do enable the peak properties of complex signals to be analyzed and amplifiers can then be properly designed to achieve a particular linearity. Not only does such a statistical analysis help to decide the *minimum* peak power an amplifier must be capable of providing, but it can also indicate the *maximum* peak power necessary. It is not economically viable to build an amplifier that provides the highest peak power (N carriers in phase) if this case rarely happens in practice. A smaller amplifier will achieve an acceptable result even if the peak power of the amplifier is exceeded only occasionally.

2.4.3 Gaussian Probability Density Function

In practice, many natural phenomena (or experiments) can be characterized by a Gaussian or normal density function, for example thermal noise. When the number of carriers, N, is large, carrier signals can also be considered as independent random variables with Gaussian probability distributions.

The normal or Gaussian probability density function with zero mean is defined as

$$f(v) = \frac{1}{\sigma \cdot \sqrt{2 \cdot \pi}} \cdot e^{-\frac{1}{2}\left(\frac{v}{\sigma}\right)^2} \tag{2.22}$$

and the *average* power can be calculated from

$$P_{av} = \int_{-\infty}^{\infty} v^2 \cdot f(v)\, dv = \sigma^2 \tag{2.23}$$

Figure 2.17 shows the effect on the shape of the amplitude density function for different values of σ—the flatter the curve, the higher the average power.

The theoretical peak power of such a signal is infinite; however, in practice the probability that a value or a voltage, exceeds, for example 3.29σ is very small (it happens only 0.1% of the time). The 0.01% limit corresponds to voltages exceeding 3.89σ and can be taken as a practical upper limit. Thus, for a single Gaussian variable the ratio of peak-to-average power with this definition is

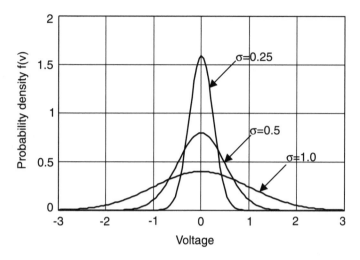

Figure 2.17 Gaussian (normal) distribution with zero mean.

$$\varepsilon = 20 \cdot \log\left(\frac{3.89 \cdot \sigma}{\sigma}\right) = 11.8 \, \text{dB}$$

A multicarrier signal consists of a number of independent random variables (carriers) and the question naturally arises as to the form of the amplitude density function in this case.

2.4.4 The Rayleigh Amplitude Density Function

The sum of a number of independent Gaussian random variables is itself a Gaussian random variable (a narrowband Gaussian signal for a multicarrier signal) with an envelope amplitude density function given by the Rayleigh distribution (Figure 2.18), that is,

$$f(v,\sigma) = \frac{v}{\sigma^2} \cdot e^{\left(\frac{-v^2}{2 \cdot \sigma^2}\right)} \tag{2.24}$$

Figure 2.18 indicates that as the amplitude tends to infinity—that is, an infinite number of carriers—the probability tends to zero. Any practical multicarrier signal has a *finite* number of carriers, and therefore the distribution function of a practical signal does not match Figure 2.18 exactly. The assumption that a signal has a Rayleigh distribution becomes less valid as the number of carriers decreases (e.g., $N < 10$) or if the carriers are correlated

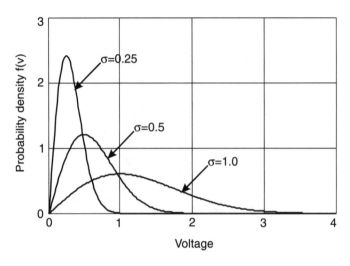

Figure 2.18 Rayleigh distribution.

(e.g., due to the modulation scheme). The Rayleigh distribution is a good and relatively simple approximation.

The average power in a signal having an amplitude distribution given by (2.24) is

$$P_{av} = \int_{-\infty}^{\infty} v^2 \cdot f(v)\, dv = 2 \cdot \sigma^2 \qquad (2.25)$$

The peak power is again theoretically infinite but for practical purposes, a value of peak-to-average ratio that gives a certain probability (e.g., 0.005%) such that the voltage level will be above a certain value is usually chosen. For example, if the peak value is chosen to be a voltage level of "v_p" volts, then the peak-to-mean ratio ψ is defined as

$$\psi = \frac{v_p}{2 \cdot \sigma^2} \qquad (2.26)$$

and the Rayleigh function can be rewritten in terms of the peak-to-mean ratio as

$$f(v) = \frac{2 \cdot \psi \cdot v}{v_p^2} \cdot e^{\frac{\left(-\psi \cdot v^2\right)}{v_p^2}} \qquad 2.27)$$

The probability that the voltage exceeds v_p can be calculated by evaluating

$$P\left(v > v_p\right) = \int_{v_p}^{\infty} f(v)\, dv \tag{2.28}$$

Solving for $v_p = 1$ gives

$$P(v > 1) = \int_{1}^{\infty} 2 \cdot \psi \cdot v \cdot e^{-\psi \cdot v^2}\, dv = e^{-\psi} \tag{2.29}$$

Figure 2.19 shows the amplitude density function for three different values of the peak-to-mean ratio ψ. Higher values of ψ imply a *lower average power* since the peak power is fixed; hence, the probability that the voltage is greater than 1 is also lower (the area under the curve for $v > 1$ is less for higher values of ψ).

Practically, it is of interest to know which value of the peak-to-mean ratio gives a certain probability that the voltage is greater than the peak value. Figure 2.20 shows that for a 0.005% probability the peak-to-mean ratio is 10 dB. This is an extremely important result since it suggests that a power amplifier should be designed with a peak-to-mean ratio of 10 dB to ensure that only 0.005% of the time a signal exceeds some maximum value (the peak power of the amplifier).

Thus, for the linear amplification of a nonconstant envelope signal, a power amplifier should be specified in terms of both the desired average

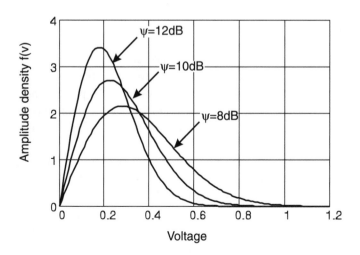

Figure 2.19 Rayleigh distribution and peak-to-mean ratio.

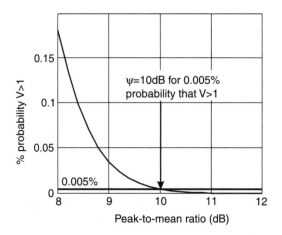

Figure 2.20 Rayleigh peak-to-mean ratio for 0.005% probability.

power and the peak power (or peak-to-mean ratio). For example, an amplifier with an average power requirement of 30W should also have a 300-W peak-power requirement if it is to amplify a signal with a Rayleigh envelope distribution (10-dB peak-to-mean ratio for a 0.005% probability that the signal exceeds the peak power). However, all signals that require linear amplification do not necessarily have a Rayleigh envelope distribution, and hence the peak-to-mean ratio can be different in different cases.

2.4.5 Effect of Baseband Filtering

The individual bits or symbols in any baseband waveform exist only for a finite duration in time; therefore, in principle at least, the power spectral density of such a waveform extends over all frequencies. To prevent interchannel interference (frequency components that fall outside the channel bandwidth), all waveforms must therefore be filtered—a process referred to as *baseband filtering.*

Different radio standards have different specifications for the amount of allowable interchannel interference, often expressed in the form of an adjacent channel power (ACP) requirement. For example, the specification for ACP in the cellular North American D-AMPS standard is 26 dB. In proposed wideband CDMA systems (e.g., WCDMA), the requirement is around 60 dB, however it can be as much as 70 dB for FM Private Mobile Radio (PMR) systems.

The effect of filtering on the baseband waveform is that the signal in the time domain spreads out such that individual bits or symbols start to overlap—a phenomena known as ISI. There is thus a dilemma: if signals are not filtered, then they occupy too much bandwidth; if they are filtered, then ISI occurs. Signals must be filtered however, and the objective is then to minimize ISI by a very careful choice of the filter characteristic (note that *Equalizers* can also be used to reduce ISI).

The filter characteristic that is most commonly used has a *raised* cosine characteristic; the frequency spectrum consists of a flat amplitude portion and a roll-off portion that has a sinusoidal form (Figure 2.21a). That is, for a data rate of R_b bits/s, bit duration T_b, and roll-off factor α, the frequency response $RC(f, \alpha)$ is given by

$$
RC(f, \alpha) = \left|
\begin{array}{ll}
T_b & \text{if } |f| \le \dfrac{R_b}{2} + \alpha \\[2ex]
T_b \cdot \cos\left[\dfrac{\pi}{4 \cdot a}\left(|f| - \dfrac{R_b}{2} + \alpha\right)\right]^2 & \text{if } \left(\dfrac{R_b}{2} - \alpha\right) \\[2ex]
& \quad < |f| \le \left(\dfrac{R_b}{2} + \alpha\right) \\[2ex]
0 & \text{if } |f| > \dfrac{R_b}{2} + \alpha
\end{array}
\right.
\qquad (2.30)
$$

As Figure 2.21(a) shows, the filter characteristic becomes more rectangular (that is, the level of adjacent channel power is reduced) as the roll-off factor $\alpha \to 0$.

In practice, to achieve the desired overall raised cosine characteristic, a root-raised cosine (RRC) filter is used in both the transmitter and receiver, that is,

$$
RRC(f, \alpha) = \left|
\begin{array}{ll}
T_b & \text{if } |f| \le \dfrac{R_b}{2} + \alpha \\[2ex]
T_b \cdot \cos\left[\dfrac{\pi}{4 \cdot a}\left(|f| - \dfrac{R_b}{2} + \alpha\right)\right] & \text{if } \left(\dfrac{R_b}{2} - \alpha\right) \\[2ex]
& \quad < |f| \le \left(\dfrac{R_b}{2} + \alpha\right) \\[2ex]
0 & \text{if } |f| > \dfrac{R_b}{2} + \alpha
\end{array}
\right.
\qquad (2.31)
$$

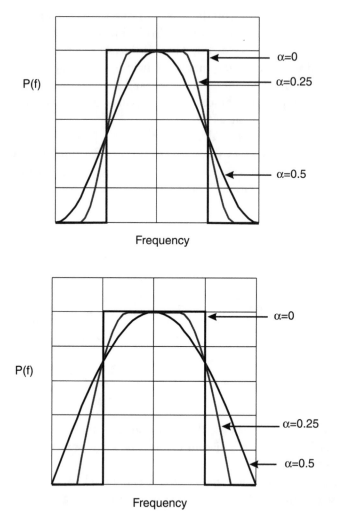

Figure 2.21 Cosine filter characteristic—frequency domain. (a) Raised cosine (b) Root raised cosine.

Figure 2.21(b) shows the frequency domain response of the RRC filter for the same cases as in Figure 2.21(a) (raised cosine). The corresponding time domain waveform (pulse shape) is shown in Figure 2.22 and is calculated from

$$\mathrm{RRC}(t, \alpha) = \frac{\cos(2 \cdot \pi \alpha \cdot t)}{1 - (4 \cdot \alpha \cdot t)^2} \cdot \frac{\sin(\pi \cdot R_b \cdot t)}{\pi \cdot R_b \cdot t} \qquad (2.32)$$

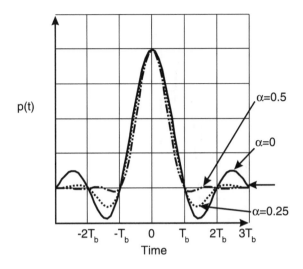

Figure 2.22 Cosine filter characteristic—time domain.

To eliminate ISI, the waveform is sampled at points where the contribution from preceding bits is at a minimum, that is, when $t = nT_b$ (n is an integer). Such a method is sensitive to timing errors, however, and it is therefore desirable that the response $RRC(t,\alpha)$ decays as rapidly as possible. As can be seen from Figure 2.22, higher values of α result in a higher/faster rate of decay (a desirable property), however, higher values also correspond to increased levels of ACP (generally undesirable).

One other disadvantage of filtering (in addition to ISI) is that the peak-to-mean ratio of a signal is increased; this is clearly undesirable in the context of power amplifier design. As the roll-off factor $\alpha \to 0$, the filter is more prone to "ringing" (a form of undamped response, for example, Figure 2.22 with $\alpha = 0$) and the peak-to-average ratio increases.

2.4.6 Linear Amplification of CDMA Signals

Consider for example, a CDMA system where each user is assigned a different binary code (the spreading waveform) but all users share the same carrier frequency. The unfiltered waveform has a typical frequency spectrum as shown in Figure 2.23(a) (single user); after filtering to limit the bandwidth of the signal, the spectrum is as shown in Figure 2.23(b). Note that the channel bandwidth depends upon the bit rate, T_b, of spreading waveform; IS95, for example, has a channel bandwidth of 1.25 MHz and WCDMA 5 MHz.

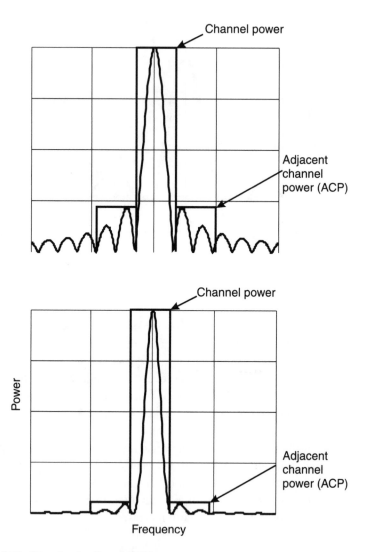

Figure 2.23 Filtered and unfiltered CDMA spectra.

Figure 2.24 shows a practical example of a filtered 1.9-GHz CDMA (IS95) signal. As previously explained, the choice of the filter roll-off factor α determines the level of ACP and this level should not be increased significantly as a result of spectral regrowth caused by nonlinearities in the power amplifier. Note also that CDMA systems may also use several carrier frequencies in the *same* transmit band (Figure 2.25); hence, distortion appears on a Δf frequency grid as well as in adjacent channels.

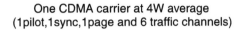

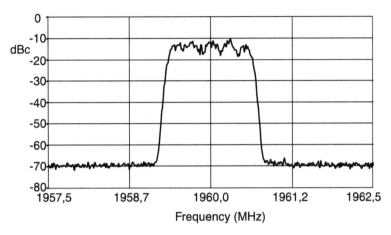

Figure 2.24 Single 1.9-GHz IS95 carrier (courtesy of Telia SA).

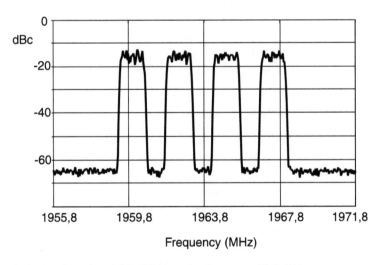

Figure 2.25 Four 750-mW 1.9-GHz CDMA carriers (courtesy of Telia SA).

The peak power of such a complex modulated signal depends upon the number of codes (the number of users) since all codes can theoretically have the same amplitude and phase for the same symbol period. For example, Figure 2.26 shows the constellation diagram for a multiuser CDMA signal using QPSK as the modulation scheme. The peak power of a CDMA signal with N

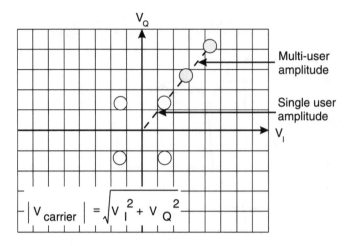

Figure 2.26 Multiuser QPSK constellation diagram.

codes is thus similar to the case of a multicarrier signal with N carriers—peaks occur when individual carriers (codes) are phase aligned. Thus, in order to specify the peak-power requirements for an amplifier, an analysis of the envelope distribution function of an N-code CDMA signal with random phases is required to calculate an acceptable peak-to-average ratio.

2.4.7 Reducing Peak-Power Requirements

In order to reduce the peak-power requirements on an amplifier, either the average power or the peak-to-average ratio must be lowered. The system link budget (Chapter 3) determines the average power, but there are several ways it can be minimized depending upon the choice of system configuration. For example, the amplifier(s) can be placed nearer to the antenna to reduce cable losses, a higher gain antenna can be used, or the number of carriers (codes) can be reduced.

When the average power has been decided upon, one possible solution to alleviate the requirements of high peak power is to deliberately reduce the peaks, that is, reduce the peak-to-average ratio, using some form of signal clipping. With a digital system this is most easily done on the baseband signal rather than the RF signal and is accomplished by limiting the maximum value of the in-phase (I) and quadrature (Q) components of the signal vectors. Although this approach has many potential advantages, a detailed system analysis must first be done to ensure that such clipping does not unduly affect system parameters such as BER or vector error.

To summarize, the peak-to-average requirement, placed on an amplifier as a result of the nonconstant envelope nature of the signals it is designed to amplify, effectively introduces a practical limit in amplifier design. High peak-power amplifiers with high peak-to-average ratios have many disadvantages including low efficiency and high cost, and every effort should be made to reduce peak-power requirements.

3

Power Amplifiers and System Design

The two basic types of RF power amplifiers are TWT amplifiers (TWTAs) and solid-state power amplifiers (SSPAs). At the lower frequency bands, for example UHF (300 MHz to 3 GHz), SSPAs predominate while TWTAs are typically used at higher microwave frequencies.

As discussed in Chapter 1, the basic building block of an SSPA is a power transistor. A single transistor can act as an amplifier, but to meet a certain gain or power output requirement, several stages containing one or more transistors are usually cascaded together. For example, Figure 3.1 shows a three-stage power amplifier; the first or input stage has high gain and low output power since the signal level is low, while the final or output stage typically has low gain but high output power. Output stages often use two or more devices in parallel to increase the available output power. The purpose of the second or driver stage is to provide sufficient input power to the output stage; if the driver is not powerful enough, then the potential high-power output will never actually be achieved. Chapter 5 shows several practical examples of different amplifier line-ups.

3.1 Amplifier Efficiency

One of the many important parameters of an amplifier is its *power conversion efficiency* (symbol η, units %). Power conversion efficiency is a measure of how effectively an amplifier converts power drawn from the dc supply to useful signal (RF) power delivered to a load, that is,

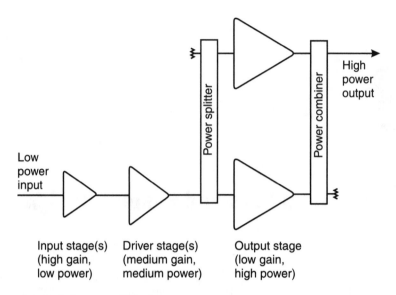

Figure 3.1 Typical power amplifier line-up.

$$\eta = \frac{P_{\text{load}}}{P_{\text{dc}}} \qquad\qquad (3.1)$$

Power that is not converted to useful signal power is dissipated as heat; and for power amplifiers that have a low efficiency, the thermal and mechanical requirements resulting from high levels of heat dissipation are often a limiting factor in a particular design.

3.2 Gain-Bandwidth Product

Amplifiers are normally designed for operation over a specific bandwidth, the transmitter band, and ideally have a gain that is constant over this bandwidth. Outside of the transmitter band, the gain response tends to drop off at both low and high frequencies (dc amplifiers are a special case). At low frequencies, components such as coupling and bypass capacitors have increasing impedances; at higher frequencies, a similar effect is caused by internal device capacitances.

The bandwidth over which the gain is within some specified limit, for example 3 dB, can be used together with the value of midband gain to define a parameter called the gain-bandwidth product. Another example of the gain-bandwidth product is the transition frequency f_T of a transistor, that is, the theoretical frequency at which the common-emitter current gain is unity.

For a given transistor, the gain-bandwidth product is often constant; hence, bandwidth and gain can be traded for each other. For example, the use of negative feedback in an amplifier allows the bandwidth to be increased at the expense of reduced gain (see Chapter 4). Constant gain over a wide bandwidth is also an important feature of feedforward amplifiers (Chapter 4); thus, gain is often reduced in favor of more broadband operation.

3.3 Classes of Amplifier Operation

The manner in which transistors are operated or biased is called the class of operation and refers to the output current waveform when an input signal is applied. The class of operation, or, more specifically, the *transistor conduction angle* (the portion of the input cycle for which the transistor conducts and an output current flows), has very important implications for power amplifiers in terms of linearity and efficiency. In any given design there is always a trade-off between linearity and efficiency; as linearity increases, efficiency decreases and vice versa.

There are many different classes of amplifier operation and bias techniques, however, the discussion here is limited primarily to Class A and Class AB amplifiers. Other classes of amplifier operation are possible—for example, Classes C, D, E, and F—however, they are not commonly used for applications requiring high linearity.

For the purposes of explanation, bipolar transistors are used as examples (in the common emitter configuration where the output current flows in the "collector"); the same principles apply to other transistor types. Note though that bipolar transistors are current-controlled devices whereas FETs are voltage-controlled and the output current flows in the "drain" rather than the collector.

3.3.1 Class A

In Class A operation (Figure 3.2(a), common emitter configuration) the transistor is biased with a dc current I_q greater than the amplitude of the signal current and therefore current (i_c) flows in the collector for the com-

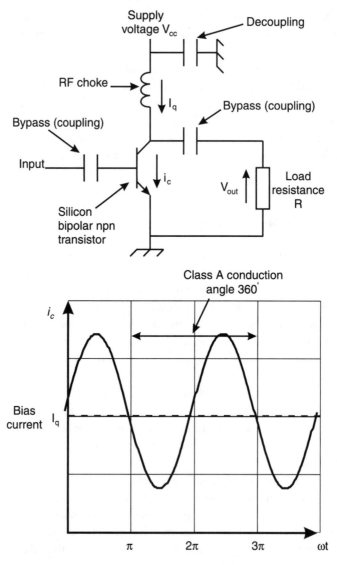

Figure 3.2 Common emitter Class A amplifier—(a) circuit diagram and (b) output current waveform.

plete duration of the input cycle; the conduction angle is thus 360 degrees. The collector output current waveform for Class A operation is shown in Figure 3.2(b).

The dc power consumption of a Class A amplifier is *independent* of the output signal amplitude and can be shown to be

$$P_{dc} = \frac{V_{cc}^2}{R} \qquad (3.2)$$

The signal power is given by

$$P_{load} = \frac{V^2}{2 \cdot R} \qquad (3.3)$$

where V is the maximum ac voltage flowing in the load, that is, $V_L = V \sin(\omega t)$. Therefore, the efficiency is (with $V \leq V_{cc}$)

$$\eta_{ClassA} = \frac{V^2}{2 \cdot V_{cc}^2} \qquad (3.4)$$

The theoretical maximum efficiency is 50% (Figure 3.3); however, in practice, the efficiency is typically less at around 30%; for signals with a high peak-to-mean ratio the efficiency becomes much lower (Section 3.4).

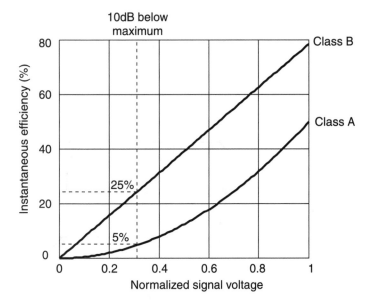

Figure 3.3 Instantaneous efficiency.

Class A amplifiers are very useful, however, when output levels are low because of their good linearity characteristics. For example, Class A amplifiers are widely used with linearization techniques such as feedforward which require a second or error amplifier that is very linear but has low output power (Chapter 4).

3.3.2 Class B

In Class B operation (Figure 3.4(a), common emitter configuration), the quiescent dc bias current is set at zero and the conduction angle is 180 degrees (Figure 3.4(b)). The output current waveform is no longer a pure sinusoid, and therefore a resonant circuit is used to regenerate the complete waveform; the resonant circuit also removes any harmonic currents. The disadvantage of using a resonant circuit is that it is inherently narrowband (it acts as a bandpass filter), and this is restrictive if operation over a wide bandwidth is required.

A solution that allows broadband operation in Class B is to use a second transistor that conducts for the negative half-cycles of the sinusoidal input (Figure 3.5). The transistors are then said to operate in *push-pull* mode

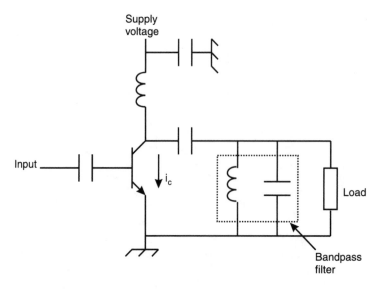

Figure 3.4　(a) Common emitter Class B narrowband amplifier—circuit diagram.

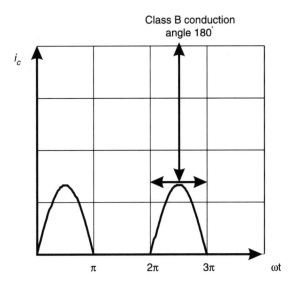

Figure 3.4 (continued) (b) Output current waveform.

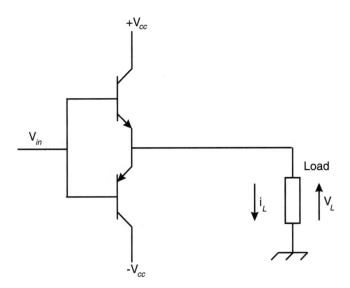

Figure 3.5 Class B push-pull configuration.

whereby one transistor *pushes* or *sources* current into the load when the input is positive and the other *pulls* or *sinks* current when the input goes negative. Note that when the input is close to zero in Class B push-pull operation, neither transistor is conducting and the output is distorted. This effect is often referred to as *crossover* or *deadband* distortion and if the goal is to make the amplifier as linear as possible, then a method must be found to reduce or completely avoid this type of distortion (see Section 3.3.3).

The dc power consumption of a Class B amplifier is proportional to the signal amplitude V and can be shown to be

$$P_{dc} = \frac{2 \cdot V_{cc} \cdot V}{\pi \cdot R} \tag{3.5}$$

The signal power is the same as for a Class A amplifier, that is,

$$P_{load} = \frac{V^2}{2 \cdot R} \tag{3.6}$$

Therefore, the efficiency is

$$\eta_{Class\ B} = \frac{\pi \cdot V}{4 \cdot V_{cc}} \tag{3.7}$$

The efficiency of a Class B amplifier is thus proportional to the output signal level and the maximum theoretical efficiency is 78% (Figure 3.3). Unlike in Class A, where the dc power consumption is constant even if there is no input signal, in Class B there is zero power consumption when the input signal is zero. Class B amplifiers are less linear than their Class A counterparts, but their much higher efficiency is a real advantage.

3.3.3 Class AB

As the name suggests, Class AB is an intermediate class between Class A and Class B. As with Class A, the transistor is biased with a nonzero dc current but the amplitude in Class AB is much less than the peak value of the output sine-wave signal. Figure 3.6 shows the output current waveform for Class AB operation; the conduction angle is greater than 180 degrees but much less than 360 degrees.

As with Class B, Class AB stages are not usually operated as single-ended stages; instead two transistors are used, one that conducts for slightly

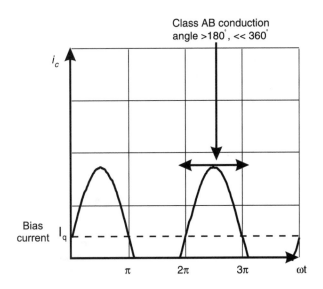

Figure 3.6 Class AB output current waveform.

longer than the positive half cycle of the input signal and the other for slightly longer than the negative half-cycle. When the input signal is close to zero, both transistors conduct and crossover distortion is thus virtually eliminated. The efficiency of Class AB amplifiers is very similar to that of Class B (Figure 3.3) except that under quiescent conditions (no input signal), a Class AB amplifier dissipates a small amount of power.

Note that the peak current demands on a device are different for different classes of operation. For example, in Class A, the peak current is determined by the bias current and is independent of the output power. In Class B and Class AB, the peak current is a function of the output power and efficiency (a more efficient amplifier has a lower peak current for a given output power).

3.3.4 Class C

The output current waveform for a Class C amplifier is shown in Figure 3.7; the conduction angle is less than 180 degrees, resulting in good efficiency but poor linearity. Class C amplifiers are therefore good for applications that require high efficiency, but their poor linearity is a significant disadvantage in many cases.

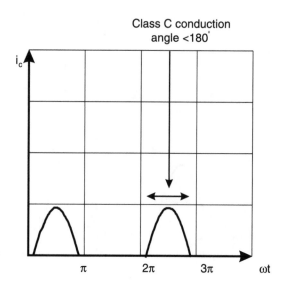

Figure 3.7 Class C output current waveform.

3.4 Efficiency and Peak-to-Mean Ratio

The instantaneous efficiency of an amplifier is most easily calculated assuming a CW signal. A narrowband signal can be thought of as a CW signal with a nonconstant envelope; for example, a multicarrier signal has an envelope that follows the Rayleigh distribution (Chapter 2). If the peak voltage is unity and the peak-to-mean ratio is ψ, the amplitude (probability) density function $f(v)$ is given by

$$f(v) = 2 \cdot \psi \cdot v \cdot e^{-\psi \cdot v^2} \tag{3.8}$$

For each envelope amplitude, v ($0 \leq v \leq 1$), the instantaneous efficiency can thus be calculated in the same way as the CW case (see (3.2) to (3.7)).

3.4.1 Class A

For a signal having an envelope that follows the Rayleigh distribution, the average (load) power of a Class A amplifier is calculated from

$$P_{av} = \frac{1}{2 \cdot R} \cdot \int_0^1 v^2 \cdot f(v) \, dv \tag{3.9}$$

A closed form of the integral can be found if the upper limit of the integral is taken as infinity rather than unity. As long as the probability of any values exceeding unity is very low, this method is valid; alternatively, numerical integration can be used to evaluate the equation as shown in (3.9).

Substituting for $f(v)$ from (3.8) and evaluating for $(0 \leq v \leq \infty)$ gives

$$P_{av} = \frac{\psi}{R} \cdot \int_0^\infty v^3 \cdot e^{-\psi \cdot v^2} dv = \frac{1}{2 \cdot \psi \cdot R} \tag{3.10}$$

The dc power consumption of a Class A amplifier with V_{cc} normalized to unity is given by

$$P_{dc} = \frac{V_{cc}^2}{R} = \frac{1}{R} \tag{3.11}$$

Therefore, the efficiency η is

$$\eta_{Class\_A} = \frac{1}{2 \cdot \psi} \tag{3.12}$$

Figure 3.8 shows the Class A efficiency as a function of the peak-to-mean ratio ψ. A signal with a 10-dB peak-to-mean ratio has an efficiency $\eta = 5\%$; and if, for example, the average output power is 30W, then the power drawn from the dc supply is 600W (570W continuously dissipated as heat).

3.4.2 Class AB

For a Class AB amplifier the average signal power is the same as for Class A (3.10), however the dc power consumption in Class AB is a function of the signal amplitude. That is, with V_{cc} normalized to unity,

$$P_{dc} = \frac{2}{\pi \cdot R} \cdot \int_0^1 v \cdot f(v) \, dv \tag{3.13}$$

Evaluating (3.13) for $(0 \leq v \leq \infty)$ to find a closed-form solution gives

$$P_{dc} = \frac{4 \cdot \psi}{\pi \cdot R} \cdot \int_0^\infty v^2 \cdot e^{-\psi v^2} dv = \frac{1}{\pi \cdot R} \cdot \sqrt{\frac{\pi}{\psi}} \tag{3.14}$$

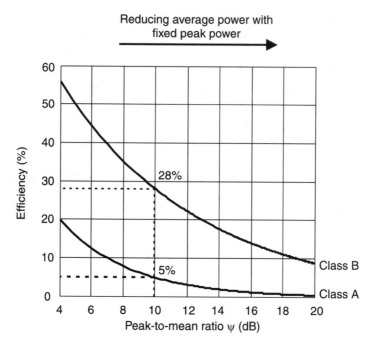

Figure 3.8 Efficiency and peak-to-mean ratio.

The efficiency is then equal to

$$\eta_{\text{Class\_AB}} = \sqrt{\frac{\pi}{4 \cdot \psi}} \qquad (3.15)$$

Figure 3.8 shows that for a Class AB amplifier and a peak-to-mean ratio of 10 dB the efficiency is 28%. Using the same example of 30-W average output power, the Class AB power consumption is $\approx$110W with 80W dissipated as heat. This is considerably less than if the amplifier was operated in Class A, but the penalty is reduced linearity.

In practice, the efficiency for both Class A and Class AB is somewhat less than the theoretical values since the output voltage saturates before the supply voltage is reached. For example, with bipolar transistors, (3.12) and (3.15) become

$$\eta_{\text{Class\_A}} = \frac{1}{2 \cdot \psi} \cdot \left(\frac{V_{cc} - V_{sat}}{V_{cc}} \right) \qquad (3.16)$$

$$\eta_{\text{Class\_AB}} = \sqrt{\frac{\pi}{4 \cdot \psi}} \cdot \left(\frac{V_{\text{cc}} - V_{\text{sat}}}{V_{\text{cc}}} \right) \tag{3.17}$$

Practical values of efficiency for Class A amplifiers at maximum power are around 30% rather than the theoretical 50%, and at 10-dB "back-off," the efficiency is around 3% rather than 5%. Class AB amplifiers at 10 dB below maximum power have a practical efficiency of $\approx$15% (see also Chapter 6).

The exact operating point of a transistor in terms of class of operation is a compromise between many factors, most notably linearity and efficiency. Once the mechanical design has been fixed (e.g., size, material and profile of the heatsink, maximum airflow, maximum safe operating temperature, and maximum ambient temperature), the linearity is largely fixed. Increasing the bias current for better linearity (closer to Class A operation) results in lower efficiency and more heat dissipation, or alternatively lowering the bias current to improve efficiency (closer to Class B) results in reduced linearity.

Note that in addition to the class of operation, the overall efficiency of an amplifier is affected by factors such as dielectric and conductor losses. For example, components such as power combiners and splitters have loss, printed circuit board materials have loss, and even signal tracks are lossy. A common part of the design procedure for power amplifiers is to first quantify any loss in the circuit, then attempt to minimize it, and finally ensure that the mechanical and thermal design is adequate under all conditions.

3.5 Compression Point and Peak Envelope Power

As previously shown in Chapter 2, the output of a non-linear amplifier compresses at high signal levels; that is, the gain drops. The output power level at which the gain has dropped by 1 dB compared to the linear value is called the 1-dB compression point, P_{1dB}. For a Class A amplifier, which has a constant gain at lower output powers, this definition is straightforward (Figure 3.9(a)); but in Class AB the gain varies and hence the concept of compression point is more difficult to use (Figure 3.9(b)). A reference level, such as the gain 10 dB below maximum power, could be defined or alternatively the rated PEP could be specified.

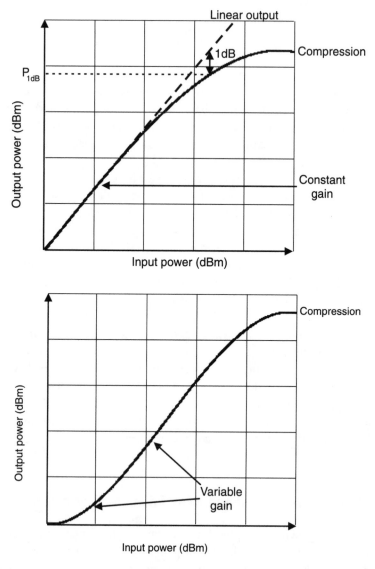

Figure 3.9 Bipolar 1-dB compression point example—Class A and Class AB.

The rated PEP of an amplifier is typically defined in terms of the maximum envelope power for a given linearity; it is common practice to specify the linearity level as −30 dBc for a two-tone signal. According to this definition, the rated PEP of an amplifier is *the maximum envelope power of a two-tone signal for which the amplifier intermodulation level is −30 dBc.*

In practice, the 1-dB compression point and rated PEP of an amplifier are approximately equal, and hence either PEP or $P_{1\,dB}$ can be used as a figure of merit. Note however, that in general, a transistor used in Class B or Class AB has a higher compression point than the same transistor used in Class A. For example, in silicon devices the compression point in Class A is related to the maximum current flowing through the device, which in turn is related to the physical size of the semiconductor material. A silicon device, for example, may have a PEP of 30W in Class AB, but in Class A the corresponding figure may be only 10W to 15W.

3.6 Intermodulation Performance

The compression point and rated PEP of an amplifier are measurements of an amplifier's linearity at maximum power; however, it is also of interest to know how linear the amplifier is at lower signal levels. For example, signals with a high peak-to-mean ratio spend a lot of time at low signal levels with the average output power typically 10 dB below the peak amplifier power to ensure that signal clipping occurs only a very small percentage of the time. An input signal is also normally defined as having a certain *dynamic range;* that is, it can vary between maximum and minimum values. For example, if the input signal has a 20-dB dynamic range and assuming that the amplifier has fixed gain, the output power will also vary over 20 dB.

The intermodulation performance of an amplifier is not only dependent upon the class of operation, it also depends upon the particular type of transistor (e.g., bipolar, MESFET, or MOSFET).

3.6.1 Silicon Bipolar

Traditionally, silicon bipolar amplifiers are used more in Class B than Class A amplifiers and Figure 3.10 shows the typical third-order intermodulation performance as a function of signal level for both cases.

In Class A, the third-order intermodulation level drops from −30 dBc to −50 dBc as the power is decreased 10 dB below compression. In Class AB, the intermodulation level is −30 dBc at rated PEP. The IM level decreases to a minimum 2 dB to 3 dB under PEP before rising again and staying close to −30 dBc until the power is sufficiently low compared to the bias point for Class A operation to be achieved.

Note that for a signal with a 10-dB peak-to-mean ratio the intermodulation performance at average power in Class AB is only −30 dBc whereas in

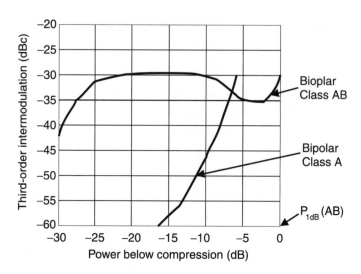

Figure 3.10 Typical bipolar third-order intermodulation performance versus signal level.

Class A it is −45 dBc to −50 dBc. A simple comparison of the efficiency (5% and 25%, respectively) illustrates why Class AB amplifiers are preferred to Class A at higher power levels.

3.6.2   Power MOSFETs

For high current capability, a MOSFET transistor should have a short channel length; however in traditional enhancement mode devices this results in a reduction of the breakdown voltage. Other structures have therefore been developed that have a short channel length and a high breakdown voltage—for example, double-diffused structures such as VMOS and LDMOS.

LDMOS transistors, in particular, are very promising for power amplifier applications at RF and have a number of advantages compared to equivalent bipolar transistors. For example:

- Better intermodulation performance;
- Good gain linearity;
- Smooth saturation (as opposed to abrupt);
- Simpler bias circuits;
- No thermal runaway for high current, Class A biasing;

- High overdrive capability;
- Good ruggedness.

One disadvantage with current LDMOS transistors, however, is that the quiescent bias point can change (drift) even if the gate voltage remains fixed. Any change in quiescent current alters the transistor bias point and can result in worse intermodulation performance.

MOSFETs are normally biased in Class AB rather than Class A and their performance in Class AB is much better than bipolar or Gallium Arsenide. The third-order intermodulation products drop to −40 dBc at 2 dB to 3 dB under rated PEP and remain at this level when the signal level is reduced (Figure 3.11). MOSFET in Class A is similar to bipolar but not as good as Gallium Arsenide (lower efficiency and the intermodulation levels drop more slowly as the signal level is reduced).

For a signal with a 10-dB peak-to-mean ratio, the intermodulation performance at average power for MOSFETs in Class AB is −40 dBc compared to −30 dBc for bipolars or Gallium Arsenide in Class AB. This is a large difference and has important consequences in the design of linear amplifiers; the more linear an amplifier is at the start, the easier it is to reach a certain linearity level using a given linearization technique. For example, if the goal is an overall intermodulation level of −55 dBc, then only a 15-dB improvement is

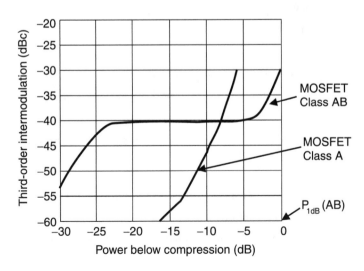

Figure 3.11 Typical MOSFET third-order intermodulation performance versus signal level.

required with MOSFETs as opposed to 25 dB with bipolars or Gallium Arsenide.

3.6.3   GaAs MESFET

Traditionally, Gallium Arsenide amplifiers are biased in Class A rather than Class AB although they are becoming more competitive in Class AB. Compared to LDMOS and bipolar in Class A, Gallium Arsenide has the best efficiency and intermodulation performance (improves more rapidly when signal power is decreased below PEP). The intermodulation performance of Class AB Gallium Arsenide amplifiers is very similar to bipolar amplifiers (Figure 3.10) in that the intermodulation level first decreases to a minimum and then remains around 30 dBc until it drops again at signal levels greater than 10 dB under PEP.

3.7   Factors Affecting Choice of Transistor

Currently, LDMOS appears to be the best choice for Class AB amplifiers while Gallium Arsenide seems preferable for Class A; however, the final choice of transistor for a particular application will depend upon many factors. First there are the performance criteria, primarily:

- Frequency of operation;
- Average power output;
- Efficiency;
- Linearity (intermodulation performance);
- Peak-power requirement (signal peak-to-mean ratio).

There is a whole host of other factors to consider when choosing a transistor, for example:

- Cost (component cost for prototype and volume production);
- Availability (number of suppliers, lead times, quantity);
- DC power requirements (availability of power modules, current capacity);
- Previous experience;

- Process reliability (consistency from one device to another—RF and dc performance);
- Device reliability (aging, MTBF);
- Production time;
- Ruggedness (e.g., handling in production, ESD precautions, maximum VSWR, overdrive capability);
- Maximum operating temperature and cooling considerations;
- Physical size and packaging.

In practice, several different kinds of transistors are used in the same amplifier since input, driver, and output stages all have varying requirements in terms of power-handling capability.

3.8 Linear Power Amplifiers

Power amplifiers can be divided into two groups:

1. Single-carrier power amplifiers (SCPAs);
2. Multicarrier power amplifiers (MCPAs).

Signals can also be broadly divided into two groups:

1. Constant envelope;
2. Nonconstant envelope.

Amplification of a single carrier with a true constant envelope presents the least problems with respect to linearity and power amplifier design and it is thus possible to use Class C or other amplifiers that have good efficiency. As soon as the envelope of the signal begins to vary, however, the linearity of the amplifier becomes much more important and a more linear amplifier than Class C (i.e., Class A or Class AB) must be used.

Although more linear than Class C, the output of Class AB and Class A amplifiers is still a distorted version of the input. Using better transistors, such as LDMOS, improves the situation but what is really required is a *truly linear amplifier;* that is, an amplifier that linearly amplifies both constant envelope and nonconstant envelope signals (single carrier or multicarrier).

Unfortunately such an amplifier does not exist; therefore, a number of techniques have been developed that attempt to "linearize" an inherently nonlinear amplifier (Chapter 4). Linear (or rather linearized) amplifiers can be used for either single- or multicarrier applications and for signals with any modulation format. Note also that linear amplifiers are not restricted to transmitter applications requiring medium or high power but can also be used as low-power (low-noise) amplifiers in receivers. A linear amplifier is effectively transparent to the carrier modulation and the number of carriers and can linearly amplify all types of signals, that is:

- A single carrier with a constant envelope;
- A single carrier with a nonconstant envelope;
- A multicarrier signal where each carrier has a constant envelope;
- A multicarrier signal where each carrier has a nonconstant envelope;
- A multicarrier signal with a mixture of constant and nonconstant envelope signals.

For example, Figure 3.12 shows a single-carrier signal (CDMA, IS95) while Figures 3.13 and 3.14 show examples of multicarrier signals (CW tones and IS95 CDMA signals). Finally, Figure 3.15 shows an example of a mixed modulation signal (FM and IS95 CDMA).

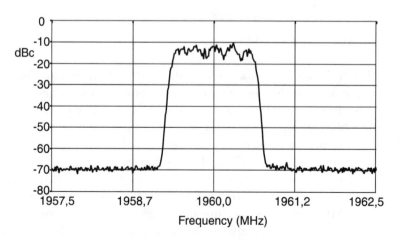

One CDMA carrier at 4W average
(1 pilot, 1 sync, 1 page and 6 traffic channels)

Figure 3.12 Single IS95 CDMA carrier (courtesy of Telia SA).

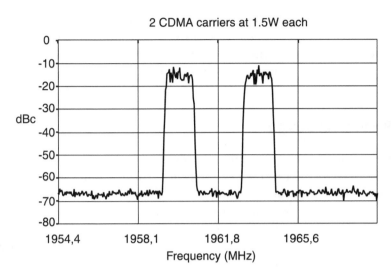

Figure 3.13 Two IS95 CDMA carriers (courtesy of Telia SA).

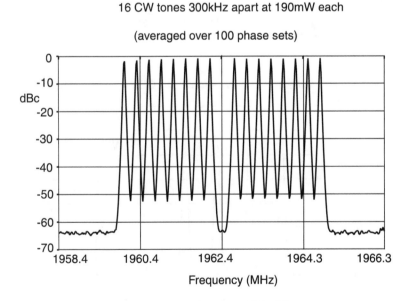

Figure 3.14 Multicarrier (16 tone) CW signal (courtesy of Telia SA).

Before looking at linearization techniques in detail, it is instructive to consider some aspects of more general *system design,* that is, how single- and multicarrier signals are generated in practice and how the subsequent development of a (linear) multicarrier power amplifier may be of benefit.

3.9   System Design

System design takes into account factors such as the choice of *radio standard* and *air interface* and uses the *link budget* to calculate and investigate the consequences of different system configurations. The primary goal of system design in terms of a power amplifier is to specify the required output power, linearity, and other performance requirements such that overall system performance and functionality are fulfilled. For commercial systems, low cost is also an important factor.

3.9.1   Radio Standards and the Air Interface

As shown in Chapter 1, Cellular, PCS, and UMTS/IMT2000 radio systems are often referred to as belonging to a particular generation, that is, first, second, or third; a particular radio system is best described by the *standard* on which it is based.

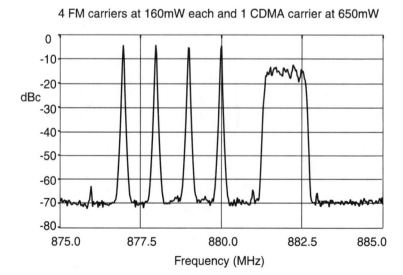

Figure 3.15 Multicarrier mixed-mode signal (courtesy of Telia SA).

A radio standard, such as NMT, D-AMPS, or WCDMA, defines the operation and interface between a number of network components; for example:

- Mobile stations;
- Base stations;
- Switching centers;
- Location registers;
- Gateways;
- Operation and maintenance centers.

In the context of radio transmission, the connection, referred to as the air interface, between a BS and an MS is of principal interest. An air interface is usually specific to a particular standard and describes in detail how communication takes place between the BS and the MS, that is:

- The RF signaling pattern (frames, subframes, and superframes);
- The division between physical and functional channels;
- Channel mapping and channel structure;
- Timeslot information;
- Timing and synchronization;
- Random access and quality control;
- Transmission output control (power regulation);
- Signaling structure.

From a power amplifier viewpoint, system design begins with the choice of radio standard and air interface since these determine the required overall radio performance, that is, necessary amplifier linearity. Different standards have different requirements in terms of, for example, the acceptable level of adjacent channel power, spurious emissions, or noise in the receiver band. One of the first tasks in system design is to study these top-level requirements and translate them into more specific requirements for subunits such as TRXs, amplifiers, and filters. In many standards there is, for example, no explicit requirement on the linearity of a multicarrier amplifier in terms of the dBc level of intermodulation products or on the peak-power requirements.

Only when the subunit requirement specifications are, in principle, agreed upon can the implementation and design stages really begin in earnest. In practice, however, it is not uncommon for new or modified requirements to appear at later stages and it is therefore important that the consequences of any change or reinterpretation of existing requirements are both understood and appreciated. For example, the mechanical, RF, and thermal design of a power amplifier is very sensitive in terms of cost and development time to changes in average- and peak-power requirements. Increasing or decreasing the average power or the signal peak-to-average ratio (e.g., due to a change in the number of carriers in an FDMA system or the number of codes in a CDMA system) can have major consequences in terms of mechanical, RF, and thermal design.

3.9.2   Power Amplifier Requirements

In order to be able to write a requirement specification for a power amplifier, the following system parameters should ideally be known in advance:

- Frequency bands (transmit and receive);
- Modulation format and carrier bandwidth;
- Signal dynamic range;
- Signal peak-to-mean ratio;
- Maximum number of carriers;
- Desired average output power;
- Power regulation requirements;
- Permissible in-band emissions;
- Permissible out-of-band emissions;
- General environmental conditions.

Specifying amplifier performance at this level can only be done after considering the *link budget,* which determines the absolute power levels throughout the system (uplink and downlink signal paths).

3.9.3   The Link Budget

As demonstrated in Chapter 1 (Figure 1.3), two distinct signal paths, commonly referred to as the *uplink* and the *downlink,* can be identified in a mobile radio system. Uplink is between the MS and the BS; downlink is from the BS to the MS. A power amplifier in a BS is part of the downlink and in an MS, part of the uplink.

Figure 1.4 showed the basic configuration of a mobile radio system, but it did not give any indication of the power levels, amount of loss, or linearity levels—this is the purpose of the link budget. That is, the link budget answers the following questions.

For the downlink:

- What is the output power of the TRXs?
- What are the losses between the amplifier and antenna?
- How directional is the BS antenna?
- What is the weakest signal the MS can successfully detect (the sensitivity of the MS)?
- For a given amplifier power, what is the maximum path loss?
- What are the acceptable linearity levels?

For the uplink:

- What is the weakest signal the BS can successfully detect (the sensitivity of the base station)?
- For a given amplifier power, what is the maximum path loss?
- What are the acceptable linearity levels?

Note that for a *balanced* system the maximum path loss on the uplink and downlink are the same. When the path losses are different—e.g., the system tolerates more loss on the downlink than on the uplink—the system is said to have a downlink overbalance.

3.9.4 Calculating Amplifier Output Power

The received signal strength for a single carrier at the input to the receiver in the MS can be calculated by adding decibel gains and losses through the BS, propagation path, and the MS, that is,

$$P_{ms\_rx} = P_{trx} + G_{amp} + G_{dpx} + G_{bsa} + G_{path} + G_{msa} \qquad (3.18)$$

where

P_{ms_rx} = MS received power (dBm)
P_{trx} = Transceiver output power (dBm)
G_{amp} = Power amplifier gain (dB)
G_{dpx} = Duplex filter gain (loss) (dB)
G_{bsa} = BS antenna gain (dB)
G_{path} = Path loss (dB)
G_{msa} = MS antenna gain (dB)

The practical lower limit (maximum path loss) for a given transmitter power is when the received power is equal to the MS sensitivity, S_{ms} (the lowest signal level for which successful detection is possible), that is,

$$S_{ms} = P_{trx} + G_{amp} + G_{dpx} + G_{bsa} + G_{path\_max} + G_{msa} \qquad (3.19)$$

Ideally the MS sensitivity should be as high as possible since for a given transmit power the path loss can be greater for higher sensitivities and the signal still detected successfully. Alternatively, for a given path loss, the transmitter power can be reduced if the sensitivity is higher. Both low transmitter power and tolerance to high path loss are desirable features of a radio system, but inevitably the final solution is a compromise since it is not practical to have a receiver that detects ultra low levels or to have an amplifier with extremely high output power.

Calculating the output power of the amplifier is an iterative procedure; that is, an output power is first chosen and then the maximum path loss is calculated for a given receiver sensitivity, TRX output power, antenna gain, and filter loss. One or more of the parameters can then be varied until an acceptable distribution of gain and loss for both the uplink and downlink is found.

Note that the term "filter loss" includes losses from cables, test devices such as directional couplers to measure forward and reverse power, and protection devices such as circulators to prevent power from flowing back into the power amplifier. Furthermore, a BS usually has several carriers and, therefore, combining losses which can be substantial must also be considered.

3.10 Combining RF Signals

Between a power amplifier and an antenna there is a succession of lossy components that turn a portion of the generated RF into heat, thereby reducing the available power at the antenna and, subsequently, the received power at an MS. Remembering that each Watt of generated RF power from an amplifier is costly (in terms of money, complexity, space, dc power consumption, cooling, and even acoustic noise), it is obviously desirable to minimize these losses as much as possible.

As explained in Chapter 1, traditional transmitter architectures for mobile radio systems are based on a number of high-power TRXs each capable of generating a single high-power modulated carrier anywhere in the transmit band. The high-power output signals from the TRXs are then combined since there is, in general, only one transmit antenna. Traditionally, the two most common methods for high-power signal combining are auto-tuned combining and hybrid combining. In both techniques each TRX is equipped with a power amplifier (an SCPA) to generate a high-power modulated carrier signal and the signals are then merged (combined) before continuing to the duplex filter and antenna. More recently, however, techniques have been developed that allow signals to be combined at low power and then one amplifier (an MCPA) is used to simultaneously amplify all carriers.

3.10.1 Auto-Tuned Combiners

Figure 3.16 shows the principle of an auto-tuned combiner. The high-power output signal from each TRX is fed into a tunable resonant cavity before being merged with the other signals. The resonant cavity, which acts as a bandpass filter and ensures good isolation between the transceiver signals, is tuned by mechanically adjusting the physical dimensions of the cavity with, for example, digitally controlled stepper motors. A tuning controller is used to automatically tune each cavity to match the TRX carrier frequencies that in turn are chosen according to local factors such as the frequency re-use pattern used by a particular BS. The combining manifold shown in Figure 3.16 is a multiple-input, single-output combiner, such as a hybrid, or a set of cavities connected in a "star" configuration.

Auto-tuned cavity combining has a number of disadvantages including large physical size, the need for mechanical adjustment of the cavities, and the requirement for a power amplifier in each transceiver; there is also the issue of combining loss to consider. Such an arrangement is also very inflexible as there are restrictions on carrier separation, and furthermore it takes a finite

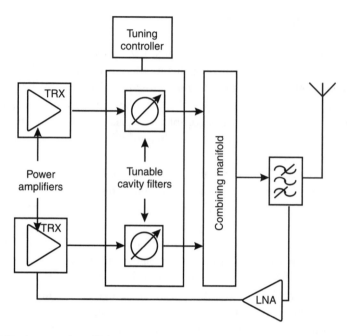

Figure 3.16 Auto-tuned combining.

time to tune to a new carrier frequency. Spectral efficiency and the ability to use schemes such as frequency hopping are thus limited when cavity-combining techniques are used.

Since the combining is done at high power, even relatively small losses will result in a large number of Watts being dissipated as heat and this further increases the demands on the power amplifiers. For example, even a loss as little as 1.5 dB (typical loss for 2 carriers) reduces the available power by 30%. As the number of carriers increases, so does the combining loss; for example, for four to six carriers the loss typically increases to ≈3 dB and 50% of the RF power turns to heat.

3.10.2 Hybrid Combiners

As shown in Figure 3.17, an alternative to combining signals at high power with resonant cavities is to use passive (lossy) hybrid combiners. Note that isolators, which also introduce loss, are required to reduce coupling and intermodulation between amplifiers.

Hybrid combining is most commonly used in BS where the number of carriers (TRXs) is low and has the advantage that the physical size is signifi-

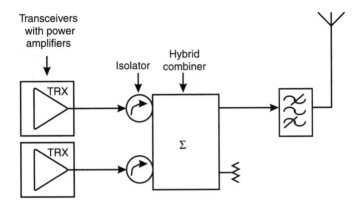

Figure 3.17 Hybrid combining, two-carrier example.

cantly smaller than for an auto-tuned combiner with the same number of carriers; the main disadvantage compared to an ATC is the higher loss. As the number of carriers increases, for example, to meet increased traffic demand, the number of combiners, and hence the loss, increases rapidly (each combiner has a typical loss of slightly more than 3 dB).

3.10.3 Multicarrier Power Amplifiers

Flexibility from a system viewpoint can be viewed as the ability to mix different modulation schemes, alter the number of carriers, and change rapidly (hop) between carrier frequencies without having to mechanically retune filters, reconfigure lossy combining networks, or add additional power amplifier stages for each new TRX. One solution that gives such flexibility is a linear multicarrier power amplifier (MCPA) (Figure 3.18). As previously explained, a multicarrier amplifier is a linear amplifier that is capable of handling a large number of signals with the same or mixed modulation formats. Individual carriers in an MCPA can be narrowband signals as in, for example, PDC (25-kHz channel bandwidth) or wideband such as WCDMA (5-MHz channel bandwidth).

By using low-power TRXs and a single high-power linear amplifier, many of the problems associated with auto-tuned or hybrid combiners are solved and many new benefits are brought to the system. For example:

- Removal of frequency spacing limitations associated with cavity combiners;

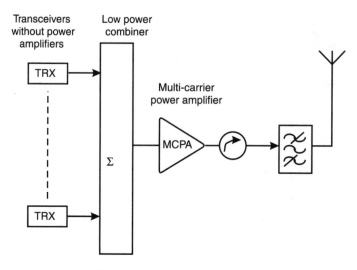

Figure 3.18 Multicarrier power amplifiers.

- Elimination of high-power combining;
- TRXs can be made without a power amplifier stage;
- Signals with different modulation schemes can use the same amplifier;
- Total output power can be increased using several MCPAs in parallel;
- Frequency hopping is possible.

For a given total average output power, the power per carrier can also be varied according to the number of TRXs (e.g., with 10W total output power it is possible to have 10 carriers @ 1W/carrier or 20 carriers @ 500 mW). Note, however, that as the number of carriers is increased, the peak-power requirement of the amplifier may increase even if the average power remains constant.

3.11 System Configurations

The quest for lower RF power requirements can lead to some interesting system design concepts, some of which are only now becoming practically realizable, for example, high-performance mast-mounted multicarrier power amplifiers in combination with high-gain adaptive array antennas. Three different system configurations and the subsequent implications for power amplifiers are now briefly considered.

3.11.1 High-Power Transceivers, High-Power SCPAs, and Medium-Gain Antenna

This is the traditional auto-tuned or hybrid combiner solution where each TRX is equipped with an SCPA (Figures 3.16 and 3.17). As previously mentioned, the major disadvantages of using this kind of approach are the high combining loss, lack of flexibility, and the need for one high-power amplifier per carrier.

3.11.2 Low-Power Transceivers, High-Power MCPAs, and Medium-Gain Antenna

This configuration has the advantages of low-power TRXs, and the flexibility that comes with multicarrier amplifiers (Figure 3.18). The major disadvantages are that the amplifier power output is still high and the efficiency is low.

Typically, several MCPAs are combined in parallel (Figure 3.19), which not only enables the output power to be increased but also gives some degree of redundancy. That is, some power still reaches the antenna even when one of the MCPAs no longer functions or is taken out of service. As Figure 3.19 shows, a high-power combiner and low-power splitter are required when using MCPAs in parallel. Note also that individual MCPAs should be matched in gain and phase. Gain matching ensures that all MCPAs have equal output power (assuming equal input power) and hence the same intermodulation performance; phase matching is necessary to avoid signal cancellation at the output of the high-power combiner and hence loss of output power.

Depending upon the number of amplifiers and mechanical arrangement, feedforward amplifiers can be combined using either a corporate feed (Figure 3.20) or serial feed structure (Figure 3.21). It is also possible to use hybrid matrix amplifiers, for example, to share the power between several feedforward modules and redirect the signals to different sectors in a cell (Figure 3.22).

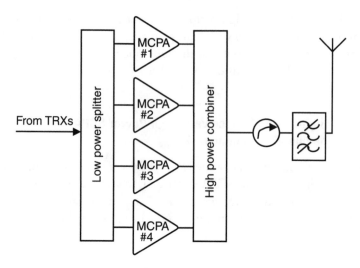

Figure 3.19 Parallel combining multicarrier power amplifiers.

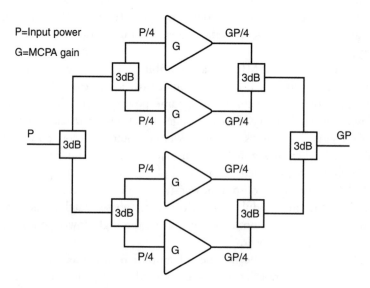

Figure 3.20 Corporate feed structure.

P=Input power

G=MCPA gain

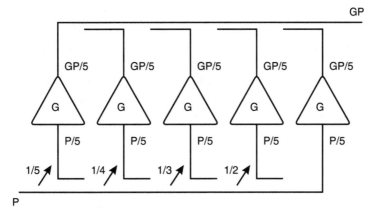

Figure 3.21 Serial feed structure.

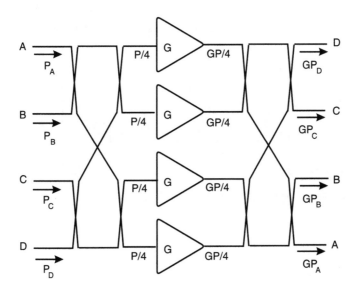

Figure 3.22 Hybrid matrix combining.

3.11.3   Low-Power Transceivers, Low-Power MCPAs, and High-Gain Antenna

In this implementation (Figure 3.23), the power output of both the TRXs and the amplifiers is low due to the use of mast-mounted amplifiers and a high-gain adaptive array antenna. This type of system configuration has many advantages, primarily the low output power of the amplifiers and the subsequent reduction in power consumption and heat dissipation. One disadvantage may be that the amplifiers are now in an outdoor environment and physically remote, thus making maintenance and replacement more difficult.

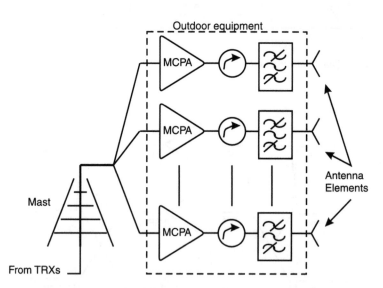

Figure 3.23   Mast-mounted multicarrier power amplifiers.

4

Linearization Techniques

A number of linearization techniques exist; feedback, predistortion, and feed-forward can all be used (separately, or sometimes in combination) to linearize an inherently nonlinear amplifier. It is also possible to generate a linear signal using the synthesis of other nonlinear signals, for example using techniques such as RF synthesis and envelope elimination and restoration.

An important property of all techniques is their linearization bandwidth, which is often described as being narrowband or wideband. The actual definition of wideband and narrowband is open to interpretation and depends upon the specific application. For example, a single carrier with a 5-MHz channel bandwidth is generally regarded as a "wideband" signal, whereas a single sideband signal (5-kHz channel bandwidth) is regarded as narrowband. In general, however, whether one particular technique or signal is narrowband or wideband is a question of definition; for example RF feedback, which can have a linearization bandwidth of several megahertz, may be regarded as either wideband or narrowband.

In the following discussion, feedback, RF synthesis, and envelope elimination and restoration are presented as examples of narrowband schemes, while predistortion and feedforward are introduced as suitable for wideband systems.

4.1 Feedback

In the late 1920s, an electronics engineer named Harold Black proposed using *feedback* as a useful circuit function and the feedback amplifier, which has since become a fundamental building block in modern electronic circuits, came into existence. A few years earlier Harold Black also proposed a technique called *feedforward* and received a patent relating to it in 1928 (U.S. Patent No. 1,686,792; see Appendix A). Feedforward, however, was largely ignored until alternative linearization methods were required for amplifiers with high delay where stability considerations precluded the use of feedback.

4.1.1 Principle of Operation

Two forms of feedback exist: positive feedback and negative feedback. Positive feedback in an amplifier is undesirable because the amplifier response can become oscillatory rather than stable; therefore, when referring to amplifiers, it is generally assumed that feedback is negative. The terms "feedback amplifier" and "negative feedback amplifier" thus become interchangeable.

In essence, negative feedback allows a designer to trade gain for some other desirable property, for example, reduced nonlinear distortion or increased bandwidth. Negative feedback can also be used to control input and output impedances, reduce the effects of noise, and make the gain of an amplifier less sensitive to variations in circuit components, for example, due to temperature effects. A limitation of negative feedback is that under certain conditions the feedback can become positive and be of sufficient magnitude to cause oscillations; there are thus stability criteria associated with a (negative) feedback amplifier (Section 4.1.2).

Figure 4.1 shows the principle of a feedback amplifier; the source and load are assumed to be perfect and do not affect the open loop gain, A, of the amplifier in any way. With no feedback applied, the open-loop transfer function (or simply the amplifier gain) is given by

$$A = \frac{V_{out}}{V_{in}} \qquad (4.1)$$

In a feedback, that is, closed-loop configuration, the output signal V_{out} is reintroduced (fed back) at the input after scaling by a factor β, the feedback factor, that is,

$$V_f = \beta \cdot V_{out} \qquad (4.2)$$

The feedback signal V_f is subtracted from the source signal V_s, generating a difference signal V_{in} that becomes the input signal to the basic amplifier

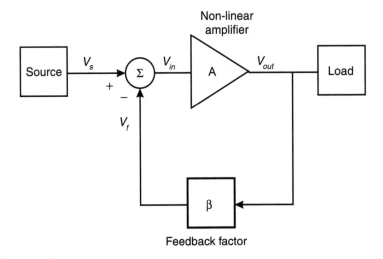

Figure 4.1 Feedback components.

$$V_{in} = V_s - V_f \tag{4.3}$$

Substituting for V_f from (4.2) gives

$$V_{in} = V_s - \beta \cdot V_{out} \tag{4.4}$$

The transfer function of the amplifier with feedback, that is, the gain V_{out} / V_s, is obtained by combining (4.1) to (4.3) to give

$$A_f = \frac{V_{out}}{V_s} = \frac{A}{1 + A\beta} \tag{4.5}$$

Note that it is the subtraction of the signals (the negative sign in (4.3)) that makes the feedback negative; negative feedback thus always acts to reduce the signal at the input to the amplifier. The feedback remains negative as long as the feedback gain, $A\beta$, is a positive quantity; that is, V_f and V_s have the same sign. If the loop gain becomes negative for some reason, then the feedback becomes positive and oscillation may occur (see Section 4.1.2).

In general, $A\beta \gg 1$ and (4.5) becomes

$$A_f = \frac{1}{\beta} \tag{4.6}$$

That is, the gain of a feedback amplifier is almost independent of the open loop gain and depends only on the feedback network, which can be chosen with a high degree of accuracy and implemented with linear passive elements. The penalty for better linearity is that the gain of the feedback amplifier is reduced by the feedback factor $(1 + A\beta)$.

4.1.2 Stability of Feedback Amplifiers

An appreciation, rather than a detailed discussion and mathematical description, of the stability criteria for feedback amplifiers is considered appropriate here and, therefore, the following analysis has been simplified considerably.

The gain of an amplifier is in general a function of frequency and, therefore, the open-loop gain A in (4.1) should more accurately be called the open loop transfer function $A(\omega)$. Similarly, the feedback network can in general contain reactive components as well as resistive components and hence the feedback factor β also becomes a function of frequency, that is, $\beta(\omega)$. The closed loop transfer function (4.5) is then

$$A_f(\omega) = \frac{1}{1 + A(\omega) \cdot \beta(\omega)} \qquad (4.7)$$

The stability of a feedback amplifier is determined by the manner in which the loop gain $A(\omega)\beta(\omega)$ varies with frequency.

4.1.3 Gain and Phase Margins

One way to investigate stability is to look at the amplitude and phase response of the loop gain in the form of a Bode plot as shown in Figure 4.2. The loop gain $A(\omega)\beta(\omega)$, which is a complex number, is represented by a magnitude and a phase, that is,

$$A(\omega) \cdot \beta(\omega) = \left| A(\omega) \cdot \beta(\omega) \right| \cdot e^{j\phi(\omega)} \qquad (4.8)$$

Note that the amplitude and phase response of the network shown in Figure 4.2 is a fourth-order one; that is, the phase approaches 360 degrees as the frequency $\omega \to \infty$.

In this particular example, when the phase angle $\phi(\omega)$ becomes 180 degrees, the magnitude $|A(\omega)\beta(\omega)|$ is less than unity (negative decibels) by an amount referred to as the gain margin. Thus, the gain margin is the difference between the value of $|A(\omega)\beta(\omega)|$ evaluated at $\omega = \omega_{180}$ and unity; gain margin

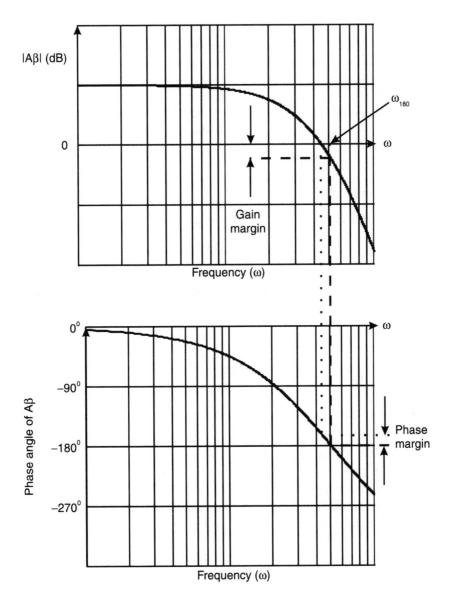

Figure 4.2 Bode plot—gain and phase margins.

represents the amount by which the loop gain can be increased while maintaining stability. In Figure 4.2, at the frequency $\omega = \omega_{180}$, although the loop gain is negative and the feedback is thus positive, the amplifier remains stable with a closed-loop gain that is higher than the open-loop gain. If, however, the magnitude of the loop gain is greater than or equal to unity (i.e., zero gain margin) at $\omega = \omega_{180}$, then the amplifier oscillates; that is, sinusoidal signals of frequency $\omega = \omega_{180}$ are present at the input and output of the amplifier.

Alternatively, the Bode plot can be used to evaluate the phase angle when $|A(\omega)\beta(\omega)| = 1$; if this angle is less than 180 degrees (e.g., as shown in Figure 4.2), then the amplifier is stable. Phase margin is thus defined as the difference between the angle at which $|A(\omega)\beta(\omega)| = 1$ and 180 degrees. In practice, feedback amplifiers are typically designed to have a phase margin of at least 45 degrees.

4.1.4 RF Feedback

Recalling from Chapter 1 that the delay of an electrical network is given by

$$\tau = \frac{\Delta\phi}{\Delta\omega} \tag{4.9}$$

The frequency $\omega = \omega_{180}$ can be calculated by setting the phase term equal to 180 degrees (π radians) and evaluating for a given delay τ, that is,

$$\Delta\omega_{180} = \frac{\pi}{\tau} \tag{4.10}$$

or, alternatively,

$$\Delta f_{180} = \frac{1}{2 \cdot \tau} \tag{4.11}$$

That is, as the delay τ increases, the bandwidth over which the amplifier remains stable ($f < f_{180}$) becomes progressively smaller.

For a feedback amplifier, the delay τ in (4.11) actually represents the delay of the amplifier plus the delay of the feedback path. In the case of the amplifier, the delay has two component parts:

1. The actual transmissive delay (baseband phase shift);
2. The phase shift introduced by poles and zeroes in the transfer function (RF phase rotation).

Increasing the power output of an amplifier—for example, by adding additional stages—increases the transmissive delay and hence reduces the bandwidth over which an amplifier is stable with negative feedback.

Consider, for example, a high-power RF amplifier where the delay is typically tens of nanoseconds; the bandwidth over which it is possible to apply negative feedback (4.11) is then of the order of megahertz. For example, if the delay of a particular amplifier is 10 ns, then the phase rotates through 180 degrees over a 50-MHz bandwidth. If the power output capability of the amplifier is subsequently increased such that the total delay becomes, for example, 25 ns, the bandwidth Δf_{180} reduces to 20 MHz. Note, however, that in practice the bandwidth is much less due to the phase margin and the delay in the feedback path.

4.1.5 Envelope Feedback

Figure 4.3 shows a typical envelope feedback system. A portion of both the input and output signals are first separated from the main path and passed through envelope detectors. The resulting amplitude modulated signals are then compared and the difference signal is passed through the loop filter and used to control an amplitude modulator. Since the feedback signal contains only amplitude information, envelope feedback can only correct for AM/AM distortion. Any phase information is lost in the detectors and therefore AM/PM distortion cannot be corrected.

One application of envelope feedback is in analog AM transmitters where the envelope frequency of the feedback signal (at any carrier frequency) is determined solely by the modulation. Since the modulation frequency is low, the system remains stable even at high delays.

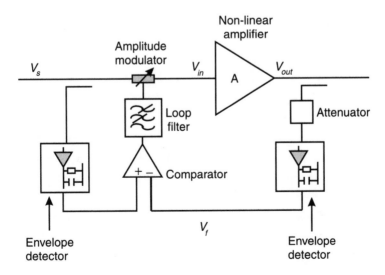

Figure 4.3 Envelope feedback.

4.1.6 Cartesian Loop Feedback

Using the in-phase and quadrature components of the input and output signals, a Cartesian loop feedback system corrects for both AM/AM and AM/PM distortion. Cartesian loop feedback is therefore superior to envelope feedback and is especially suited for linear modulation schemes that use both amplitude and phase to convey information.

Figure 4.4 shows an example of the open- and closed-loop response of a Cartesian feedback transmitter, while Figure 4.5 shows a typical Cartesian loop architecture. The inclusion of an IQ-modulator and demodulator gives the Cartesian loop the functionality of a transmitter and the feedback signals are amplitude signals, I_f and Q_f (Cartesian signal representation). Cartesian loops can be built with single or double loops (see Section 4.6) and can be implemented using ASIC technology, making them ideally suited for use in handsets. For reasons previously discussed, however, Cartesian feedback is restricted to narrowband applications such as a single 25-kHz carrier or several 5-kHz carriers.

4.1.7 Polar Loop Feedback

Like Cartesian feedback, polar loop feedback corrects for both AM/AM and AM/PM distortion. As the name suggests, polar loop uses the polar representation of a complex signal—that is, the amplitude and phase—rather than

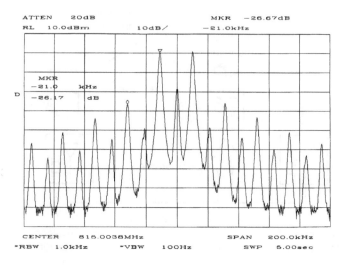

Figure 4.4 Cartesian loop example (a) open-loop (no feedback) (courtesy of Toracomm Ltd.).

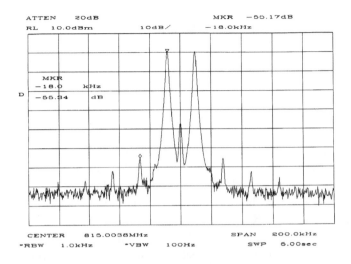

Figure 4.4 (continued) (b) Closed-loop (with feedback).

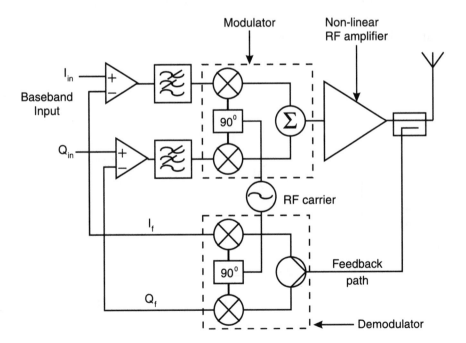

Figure 4.5 Cartesian feedback transmitter.

Cartesian equivalent (I and Q components). One disadvantage with the polar representation of complex signals is that the phase component can have a significantly wider bandwidth than either the I or Q component in a Cartesian representation of the same signal.

4.2 RF Synthesis

Linear amplification with nonlinear components (LINC) and combined analog locked loop universal modulator (CALLUM) are examples of narrowband linearization schemes that use RF synthesis techniques.

Figure 4.6 shows the principle of a LINC transmitter whereby voltage controlled oscillators (VCOs) are used to produce two phase modulated signals. The output signal amplitude (after combining the amplifier outputs) is a function of the phase in the two signal paths and can vary from zero (opposite phase) to maximum (phase alignment). Disadvantages with the LINC approach include the fact that it is an open-loop system and that the two signal paths must be very accurately matched. The bandwidth of the phase modulated signals can also be large and the vector combining can also be a problem. CALLUM solves some of these problems by adding feedback, however, neither LINC or CALLUM have yet been commercially proven.

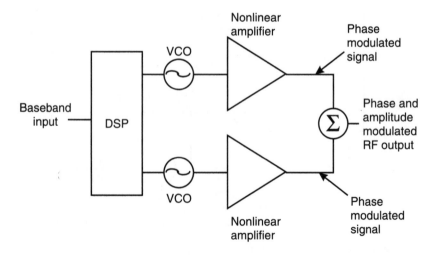

Figure 4.6 Linear amplification using nonlinear components (LINC).

4.3 Envelope Elimination and Restoration

Another narrowband synthesis method is envelope elimination and restoration, also referred to as the Kahn method. Figure 4.7 shows the principle of operation whereby the input signal is first split into two paths, one going to a limiter, the other to an envelope detector. The limiter removes the amplitude modulation component of the input signal, leaving a constant envelope phase modulated signal. The output of the envelope detector is an amplitude modulated signal, that is, a nonconstant envelope signal. An efficient but nonlinear RF amplifier (e.g., a Class C amplifier) is used to amplify the phase modulated signal, while the amplitude modulated signal is amplified using a low-frequency amplifier. The amplified, amplitude modulated signal is then used to remodulate the amplified phase modulated signal. The envelope wave shape of the high-power output signal after the modulator stage is thus the same as that of the input signal, as is the phase modulation component. Note that to ensure linear output, the time relationships between amplitude and phase modulation signals must also be properly maintained.

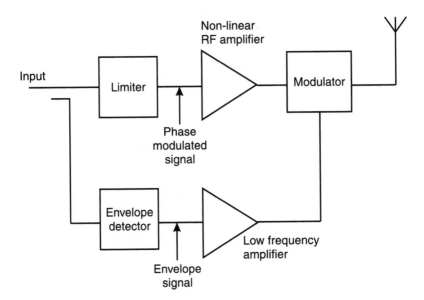

Figure 4.7 Envelope elimination and restoration.

4.4 Predistortion

As discussed in Chapter 2, the nonlinear output of an amplifier can be represented by a polynomial, that is,

$$V_{out}(t) = G_1 \cdot V_{in}(t) + G_2 \cdot V_{in}^2(t) + G_3 \cdot V_{in}^3(t) + \cdots + G_n \cdot V_{in}^n(t) \quad (4.12)$$

The amount of AM/AM and AM/PM distortion introduced by the amplifier is a function of the signal level and the relative contributions of the amplifier coefficients $G_1 \ldots G_n$. If the coefficients are known (e.g., from measurement and/or simulation), then the amplifier distortion can be compensated by introducing a nonlinearity, which when cascaded with the amplifier provides linear gain; this is the principle of predistortion. Figure 4.8 shows a typical implementation whereby a predistortion circuit operating at low power introduces a nonlinearity before the amplifier; the optimal nonlinearity is the inverse of the amplifier transfer characteristic.

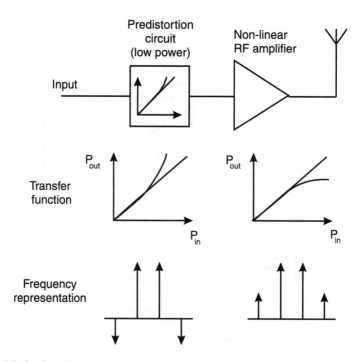

Figure 4.8 Predistortion.

The predistortion function can be implemented at baseband (e.g., adaptive baseband predistortion) or at IF/RF (intermediate or radio frequencies). Predistortion can correct for both AM/AM and AM/PM distortion and is not restricted in bandwidth since there is no inherent feedback path (note, however, that baseband and DSP predistortion *are* bandwidth limited). Fixed or adaptive predistortion schemes can be used, with the latter being able to compensate for changes in the amplifier characteristic over time, for example, due to temperature.

Disadvantages of predistortion include the fact that predistorters are normally optimized for a specific power level and that they can typically only provide limited reduction in distortion, normally only of third-order distortion products. Unlike many other linearization techniques, however, predistortion does not significantly reduce the efficiency of an amplifier. For example, when used in combination with feedforward linearization, even modest improvements in third-order distortion levels can lead to improved overall system efficiency (Chapter 5).

4.5 Feedforward

Feedback compares the output of a nonlinear amplifier with its input and uses the same amplifier to amplify the difference signal; in contrast, feedforward uses two amplifiers and there is a continuous forward signal flow, that is, no feedback path. The lack of an inherent feedback path means that feedforward is unconditionally stable and allows operation over theoretically unrestricted bandwidth. Feedforward is thus classed as a wideband linearization technique unlike feedback, which is inherently more narrowband.

4.5.1 Principle of Operation

Figure 4.9 shows a simplified representation of a feedforward amplifier. The input is first split into two paths by a power splitter (usually a directional coupler) with one path going to the nonlinear main amplifier and the other going to a delay element. A portion of the distorted main amplifier output is separated from the main amplifier path using a second coupler as a power divider and after appropriate scaling subtracted from the delayed feedforward input. The resulting error signal, ideally containing only distortion components, is then amplified by an error amplifier before being subtracted from a delayed version of the main amplifier output, thus canceling the distortion components in the main path. For example, Figure 4.10 illustrates how the

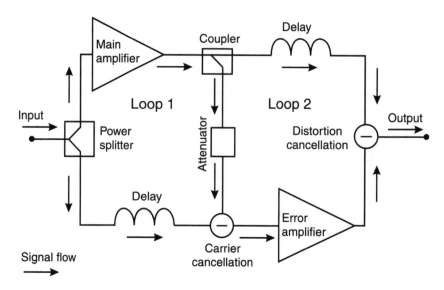

Figure 4.9 Feedforward components.

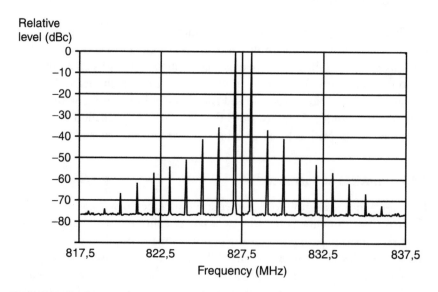

Figure 4.10 Two-tone distortion (a) before feedforward correction.

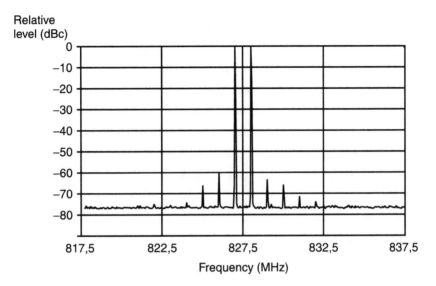

Figure 4.10 (continued) (b) After feedforward correction.

distortion that is added to a two-tone signal by a nonlinear amplifier is canceled using feedforward; the distortion is isolated in the first feedforward loop (Loop 1) and canceled in the second (Loop 2). Figures 4.10(a,b) show the frequency spectrum before and after cancellation; Figure 4.11 shows the error signal after carrier cancellation in Loop 1.

4.5.2 Input-Output Signal Linearity

In practice, any feedforward input signal (transceiver output signal) has a finite level of distortion and a feedforward amplifier cannot reduce this level, but only maintain or increase it. The output of a feedforward amplifier is therefore only completely distortion-free when the input signal contains no distortion (an ideal case and not possible in practice). For example, if a particular input signal has a distortion level of −60 dBc, then the output signal has a linearity not better than −60 dBc.

An ideal feedforward amplifier does not add any distortion to an input signal, it is a true linear amplifier. In practice, however, feedforward amplifiers inevitably introduce some distortion and, strictly speaking, the output is no longer linear. Terms such as "very-linear" and "ultra-linear" are therefore sometimes used to emphasize that the practical linearity of a particular amplifier is close to but not equal to ideal.

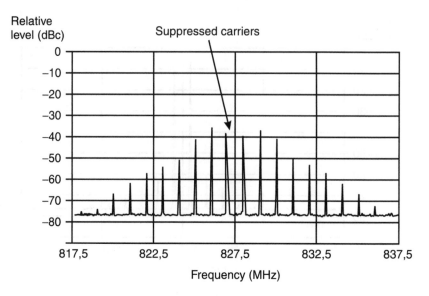

Figure 4.11 Error signal with suppressed carriers.

4.5.3 Multicarrier Input and Noise Performance

As previously noted, for an input signal that consists of many carriers with uniform carrier spacing Δf, the intermodulation products also appear on a Δf grid and cannot be removed by filtering. Removing distortion components that appear at carrier frequencies does not, however, present a problem in feedforward since the distortion is isolated in the first loop and canceled at the output of the second. The carriers at the output are unaffected as long as Loop 1 cancellation is maintained.

Another very useful property of feedforward is that in addition to canceling intermodulation distortion, the second loop also cancels distortion in the form of noise added in the main amplifier path. Feedforward amplifiers with their attendant low-noise figure are therefore particularly suited to applications where signal levels are low, such as cable systems. Indeed one of the first large-scale commercial applications for feedforward was cable TV systems.

4.5.4 Signal Cancellation

Feedforward depends upon the successful isolation of an error signal and the removal of distortion components, both of which involve signal cancellation over a band of frequencies. Mathematically, signal cancellation at a single frequency is represented by the subtraction of two signals with equal amplitude; the resultant has a magnitude equal to zero or $-\infty$ dB. In practice, cancellation is achieved by the vector addition of signals, or more specifically voltages, with equal amplitude but opposite phase.

Perfect broadband cancellation, that is, vector cancellation over a band of frequencies, occurs only when signals have:

- Equal amplitude;
- 180-degree phase difference;
- Equal delay.

Equal amplitude and opposite phase are sufficient for cancellation at a single frequency, however, the requirement of equal delay is necessary for broadband signal cancellation (hence the inclusion of delay elements in feedforward). Effects such as frequency-dependent amplitude and phase ripple also affect broadband signal cancellation; see Chapter 5 for a detailed analysis and examples.

4.5.5 Gain and Phase Adjustment

The gain and phase through the various feedforward signal paths (Figure 4.9) is, in general, a function of many parameters, for example, signal level and temperature; therefore, some form of gain and phase adjustment is required to ensure that the signals are always matched (Section 4.2.4).

Gain and phase adjustment of an RF signal can be done in several ways, including using separate amplitude and phase control networks or IQ networks; the circuits themselves can be implemented as discrete components or as ASICS. Figure 4.12, for example, shows an IQ control circuit that uses two dc voltages to control the gain and phase of the RF signal. Important parameters of such gain and phase adjustment networks include:

- The dynamic range over which the amplitude can be varied;
- The dynamic range over which the phase can be varied;
- Linearity and power-handling capability;
- Noise figure;
- Frequency response (i.e., amplitude and phase ripple);
- Physical size and cost.

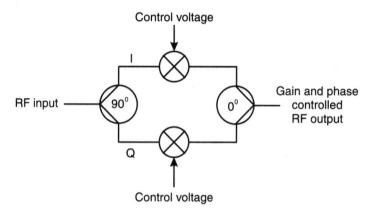

Figure 4.12 Gain/phase control network—IQ modulator example.

For example, a certain gain and phase adjustment network may have a dynamic range of 20 dB with 360-degree phase control, a 1-dB compression point power of 10 dBm, and a noise figure of 10 dB. Another network may only have 90-degree phase control, a compression point power of 0 dBm, and a noise figure of 20 dB.

The position of gain and phase control networks in a feedforward amplifier is also important. As shown in Chapter 5, to ensure that the overall gain of the feedforward system is constant, gain and phase control networks should be placed in series with the main and error amplifiers rather than the delay lines (Figure 4.13).

4.5.6 Gain and Phase Accuracy

Perfect signal cancellation implies that the resultant vector, that is, the suppressed signal, has a magnitude of $-\infty$ dB relative to the unsuppressed signal. For practical purposes, however, it is of interest to know how closely signals need to be matched for a certain finite suppression, for example, 30 dB, 40 dB, or any other value.

Figure 4.14 shows the vector addition of two voltages: a reference signal with amplitude 1 and phase 0 degrees, and a nonideal canceling signal with amplitude $1 + \delta A$ and phase 180 degrees $+ \phi$. Using the cosine rule, the magnitude of the resultant vector r, which has the same frequency as the canceling signals, can be calculated from

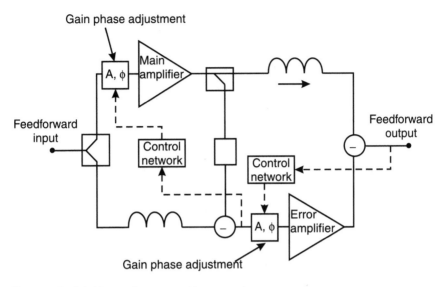

Figure 4.13 Gain/phase adjustment and loop control.

$$r^2 = (1 + \delta A)^2 + 1 - 2 \cdot (1 + \delta A) \cdot \cos(\phi) \qquad (4.13)$$

The level of the canceled or suppressed signal is thus a function of two variables, the amplitude mismatch δa and the phase mismatch ϕ. Rewriting (4.13) in terms of the suppression R (dB) and amplitude mismatch ΔA (dB) gives

$$R(\text{dB}) = 10 \cdot \log\left(\left|10^{\frac{\Delta A(\text{dB})}{10^{10}}} + 1 - 2 \cdot 10^{\frac{\Delta A(\text{dB})}{20}} \cdot \cos(\phi)\right|\right) \qquad (4.14)$$

The suppression R in (4.14) can be presented graphically in several ways:

- The phase can be varied for a fixed amplitude mismatch.
- The amplitude can be varied for a fixed phase mismatch.
- The range of amplitude and phase mismatch, for a fixed-loop suppression, can be calculated.

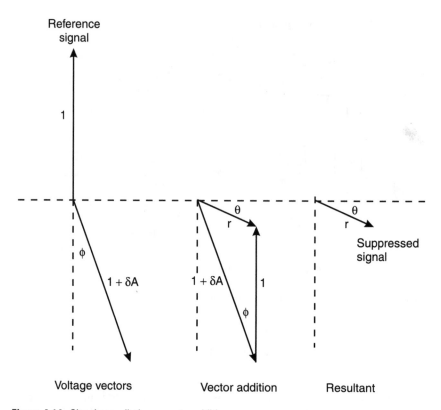

Figure 4.14 Signal cancellation—vector addition.

For example, to calculate the range of amplitude and phase mismatch for a fixed loop suppression, (4.14) can be re-arranged in terms of ΔA to give a quadratic equation with solutions:

$$\Delta A = 20 \log\left(\cos(\phi) \pm \sqrt{\cos(\phi)^2 - 1 + 10^{\frac{R}{10}}} \right) \tag{4.15}$$

Equation (4.15) is shown in Figure 4.15(a) (four quadrants) and Figure 4.15(b) (one quadrant) for different values of suppression. As the suppression

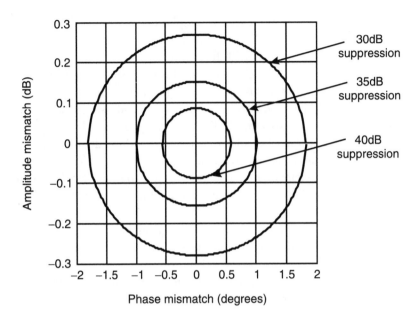

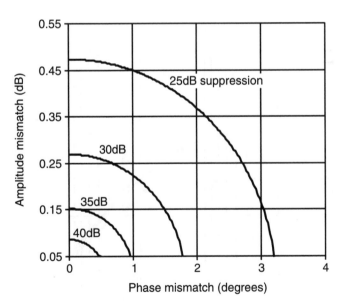

Figure 4.15 Loop suppression and gain/phase matching requirements.

level approaches −∞ dB, the range of amplitude and phase mismatch becomes progressively smaller. Typical suppression levels for practical feedforward amplifiers are in the 20- to 40-dB range.

Alternatively, Figure 4.16 shows the effects on suppression of varying the phase mismatch when ΔA is fixed. The signal amplitude and phase must be matched to within ≈0.5 dB and 0.5 degrees for 25-dB signal cancellation, and ≈0.1 dB and 0.1 degrees for 40-dB signal cancellation. Matching signals to within a tenth of a decibel and a degree is very difficult in practice and becomes even harder when the same accuracy must be maintained over a wide bandwidth. In general, it is therefore desirable to minimize the gain and phase accuracy requirements and the bandwidth over which signals are canceled.

For example, matching requirements for distortion cancellation at the output of a feedforward amplifier are dependent upon the linearity of the main amplifier and can be made less stringent if the linearity of the amplifier is improved. However, as discussed in Chapter 3, there is a trade-off between amplifier linearity and efficiency—increased linearity can normally only be

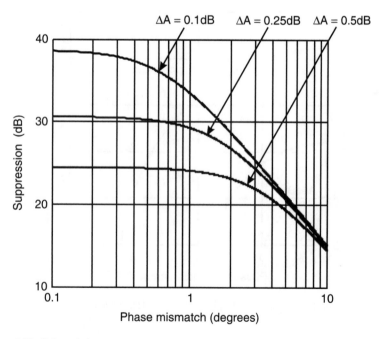

Figure 4.16 Gain and phase accuracy.

obtained at the expense of lower efficiency. One potential solution for reducing the matching requirements without adversely affecting efficiency is to use transistors that have both good linearity and reasonable efficiency, for example, MOSFETs.

4.5.7 Loop Cancellation Bandwidths

In the first feedforward loop, a time-delayed version of the input is compared to a sample of the distorted main amplifier output and, ideally, the resulting error signal consists of only the amplifier distortion products. The signals being canceled are those appearing at the feedforward input, that is, the nominally distortion-free carriers. The cancellation bandwidth in the first loop is thus equal to the maximum bandwidth of the input signal, that is, the system transmitter bandwidth. For example, if the transmitter bandwidth is 20 MHz, then the first loop cancellation bandwidth is also 20 MHz.

In the second feedforward loop, distortion components that appear at frequencies outside the transmitter band are canceled and therefore the cancellation bandwidth is larger than in the first loop. The second loop cancellation bandwidth can be defined as the bandwidth over which significant distortion appears and it is dependent upon the exact nature of the nonlinearities in the main amplifier. For example, using the same transmitter bandwidth as in the previous example, that is 20 MHz, a typical class AB amplifier may produce significant distortion over 100 MHz. The required cancellation bandwidth in Loop 2 is then five times the cancellation bandwidth in Loop 1.

For feedforward schemes that use additional cancellation loops, the same principles apply when determining loop cancellation bandwidths:

- When carriers are being canceled to generate an error signal, the loop cancellation bandwidth is equal to the transmitter bandwidth or maximum carrier spacing.

- When distortion components are being canceled to provide a linear output signal, the loop cancellation bandwidth is equal to the bandwidth over which significant distortion appears.

For example, Section 4.3 describes a dual-loop feedforward amplifier that has two carrier cancellation loops and two distortion cancellation loops.

4.5.8 Loop Suppression

Having identified the bandwidth over which cancellation is required, that is, the cancellation bandwidth, the next step is to determine the required loop suppression and hence the gain and phase matching requirements.

The primary factors in determining the required loop suppression are the linearity of the main amplifier and the desired output linearity. For example, a typical Class AB bipolar or GaAs amplifier has an intermodulation performance of ≈-30 dBc and, if the required output distortion level is -70 dBc, then 40 dB of distortion cancellation must be provided. Alternatively, a Class AB, LDMOS amplifier with an intermodulation performance of -40 dBc only requires a distortion cancellation of 30 dB.

When carriers rather than distortion products are canceled, the loop suppression must be sufficient such that the residual carrier signal, now part of the error signal in the distortion cancellation loop, does not affect the carriers at the output. Note that any residual carrier reaching the distortion cancellation point is added as a vector quantity to the signal from the main amplifier path. Whether the residual carrier signal adds or subtracts to the main signal therefore depends on the relative phases of the two signals.

In practice, 30 dB of suppression is considered sufficient in carrier cancellation loops; however, there are other factors to consider such as the loading on the error amplifier and the effects on loop control schemes.

4.5.9 Loop Control

Feedforward allows operation over a theoretically unrestricted bandwidth; however, with such an open-loop configuration, changes in device characteristics, for example with respect to time, temperature, voltage, and signal level, are not compensated. A feedforward system cannot therefore monitor its own performance unless some kind of automatic control scheme is implemented whereby the gain and phase are continuously adjusted to achieve the best signal cancellation and output linearity.

Figure 4.13 illustrates the principle of loop control. The first step is to obtain information on how well a particular loop is balanced and thereafter to feed back the information in order to adjust the gain and phase until the required loop suppression is achieved. Thus with automatic control, the feedforward loops are continuously monitored and the required output linearity is maintained as a function of time, temperature, supply voltage, and signal level. Digital and/or analog techniques can be used for loop control, and a more detailed discussion is given in Chapter 6. It should be noted, however, that the development of control networks for gain and phase adjustment is a major part of feedforward design.

4.6 Dual-Loop Feedforward

Feedforward is a technique used to linearize a nonlinear (main) amplifier. The feedforward system is itself just an amplifier, albeit more linear than the original main amplifier, and it can therefore be further linearized by a second application of feedforward. That is, a feedforward amplifier can be used as the main amplifier in a second feedforward amplifier—such a system is called dual-loop feedforward and Figure 4.17 shows the principle of operation.

Dual-loop feedforward allows both average and peak intermodulation to be improved significantly compared to single-loop feedforward at the ex-

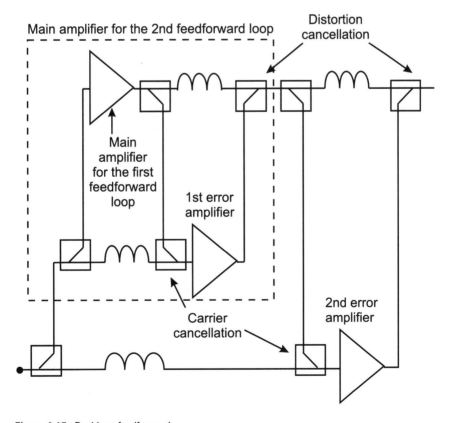

Figure 4.17 Dual-loop feedforward.

pense of increased complexity. As Figure 4.17 shows, in a dual-loop feedforward system there are two carrier cancellation loops and two distortion cancellation loops, all of which require some form of loop control. Note that, in practice, the two error amplifiers are often chosen to be identical.

Dual-loop feedforward is usually chosen as the linearization technique for applications requiring very low levels of distortion (e.g., less than −70 dBc) over a wide bandwidth and wide dynamic range; in most other cases, single-loop feedforward, or a combination of single-loop feedforward and predistortion, is used.

5

Feedforward Analysis

The low efficiency of feedforward amplifiers (in general <10% and in practice sometimes <5%) is one reason why other linearization techniques have been favored in preference to feedforward. Other reasons include the open-loop nature of feedforward and the need to define circuit characteristics to within a fraction of a decibel over the frequency band of interest. Despite these disadvantages, however, feedforward offers many system benefits including the ability to linearly amplify both constant and nonconstant envelope signals (e.g., multicarrier and linearly modulated signals). Furthermore, feedforward allows ultra-linear operation over a wide bandwidth.

Important parameters of feedforward amplifiers include gain, input and output return losses, reverse intermodulation, noise figure, linearity and bandwidth of operation (broadband signal cancellation), average and peak-power requirements of the main and error amplifiers, and efficiency.

5.1 Feedforward Gain

The purpose of this analysis is to show that a balanced feedforward amplifier has a linear gain, which is independent of nonlinearities in the main and error amplifiers. A simple formula for calculating feedforward gain is also derived.

Figure 5.1 shows the gains and losses of the individual components in a feedforward system. Couplers have insertion loss α and coupling factor c (Chapter 1), amplifiers have gain g, and delay lines and attenuators have loss l. In this simplified representation, an amplifier gain actually represents the

139

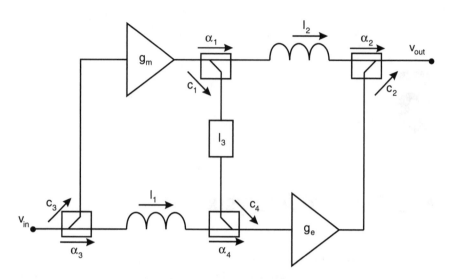

Figure 5.1 Feedforward gain analysis.

overall gain of all components in the path and includes, for example, the gain or more typically, the loss of the amplitude and phase adjustment networks.

Feedforward gain is defined as

$$g_{ff} = \frac{V_{out}}{V_{in}} \tag{5.1}$$

The output voltage, V_{out}, is the sum of two voltages, one from the main amplifier path and the other from the error amplifier, that is,

$$\begin{aligned}
V_{out} &= c_3 \cdot g_m \cdot \alpha_1 \cdot l_2 \cdot \alpha_2 \cdot V_{in} \cdots \\
&+ c_2 \cdot g_e \cdot (\alpha_3 \cdot l_1 \cdot \alpha_4 - c_3 \cdot g_m \cdot c_1 \cdot l_3 \cdot c_4) \cdot V_{in}
\end{aligned} \tag{5.2}$$

Substituting in (5.1) gives

$$g_{ff} = c_3 \cdot g_m \cdot \alpha_1 \cdot l_2 \cdot \alpha_2 + c_2 \cdot g_e \cdot (\alpha_3 \cdot l_1 \cdot \alpha_4 - c_3 \cdot g_m \cdot c_1 \cdot l_3 \cdot c_4) \tag{5.3}$$

When the first loop is perfectly balanced,

$$\alpha_3 \cdot l_1 \cdot \alpha_4 = c_3 \cdot g_m \cdot c_1 \cdot l_3 \cdot c_4$$

$$g_m = \frac{\alpha_3 \cdot l_1 \cdot \alpha_4}{c_3 \cdot c_1 \cdot l_3 \cdot c_4} \tag{5.4}$$

The gain and phase of the main amplifier are thus constant and there is no distortion at the output. Substituting for g_m in (5.3) gives

$$g_{ff} = \frac{\alpha_1 \cdot \alpha_2 \cdot \alpha_3 \cdot \alpha_4 \; l_1 \cdot l_2}{c_1 \cdot c_4 \cdot l_3} \qquad (5.5)$$

That is, the feedforward gain is a function only of the loss of the passive elements (delay lines, attenuators, and couplers) and is independent of any nonlinearities in the main amplifier.

Alternatively, when the second loop is perfectly balanced,

$$\alpha_1 \cdot l_2 \cdot \alpha_2 = c_1 \cdot l_3 \cdot c_4 \cdot g_e \cdot c_2$$
$$g_e = \frac{\alpha_1 \cdot \alpha_2 \cdot l_2}{c_1 \cdot c_2 \cdot c_4 \cdot l_3} \qquad (5.6)$$

Thus, the error amplifier has constant gain and phase, that is, no distortion at the output. Furthermore, substituting for g_e in (5.3) gives

$$g_{ff} = \frac{\alpha_1 \cdot \alpha_2 \cdot \alpha_3 \cdot \alpha_4 \cdot l_1 \cdot l_2}{c_1 \cdot c_4 \cdot l_3} \qquad (5.7)$$

Note that (5.7) is identical to (5.5); thus, when both feedforward loops are balanced, the feedforward gain is independent of any nonlinearities in the main and error amplifiers and the output contains no distortion. A balanced feedforward amplifier therefore behaves as a linear amplifier.

Feedforward gain is more easily calculated when all quantities are expressed in units of decibels; (5.7) can then be rewritten as

$$G_{ff} = A_1 + A_2 + A_3 + A_4 + L_1 + L_2 - C_1 - C_4 - L_3 \qquad (5.8)$$

Figure 5.2 shows an example of a feedforward amplifier with a gain of 30 dB. Note that the main and error amplifiers must have a higher gain than the feedforward gain in order to balance the loops. For example, the main amplifier has to have sufficient gain to overcome the losses in its path, that is,

$$G_m = G_{ff} - C_3 - A_1 - L_2 - A_2 \qquad (5.9)$$

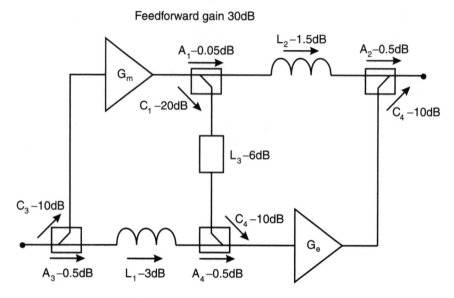

Figure 5.2 Feedforward gain example.

Similarly, the error amplifier has to have a higher gain than the feedforward gain, that is,

$$G_e = G_{ff} - A_3 - L_1 - A_4 - C_2 \tag{5.10}$$

For the example shown in Figure 5.2, the gain of the main and error amplifiers is 42 dB and 44 dB, respectively, compared to a feedforward gain of 30 dB. Furthermore, the loop gain and phase adjustment networks placed before the main and error amplifiers are typically lossy and increase the gain demands on the amplifiers. In practice, providing extra gain at low power is reasonably simple; however, it is undesirable to have too many high gain stages due to the risks of instability and low-level signal coupling.

For fixed coupler values and fixed delay line loss, the feedforward gain can be adjusted by altering the attenuation L_3. Increasing the amount of attenuation results in higher feedforward gain; however, the gain of the main amplifier path must also be increased to ensure that the main amplifier and, hence, feedforward output power remain constant. When feedforward amplifiers are being built in a production environment, the attenuator L_3 can be used to trim the feedforward gain to within a specified tolerance. It can also be used to compensate for temperature-dependent delay line loss.

5.1.1 Gain Versus Signal Level, Frequency, and Temperature

Ideally, in order to maintain constant output power (for a given input signal level), the gain of a feedforward amplifier should be constant over the pass-band of the amplifier and independent of temperature. Furthermore, the gain should not vary with signal level.

In practice, the gain of a feedforward amplifier varies with temperature, frequency, and signal level. The goal is to minimize these variations and hence minimize the changes in output power. The amount of gain variation is essentially related to how effective the loop control networks are at maintaining loop balance under all conditions (e.g., over temperature, signal level, and bandwidth). Figure 5.3, for example, shows the gain-frequency characteristic for a 1.9-GHz, 30-W feedforward amplifier and Figure 5.4 shows the gain-temperature characteristic for the same amplifier. The gain flatness is ±0.2 dB over a 60-MHz bandwidth and <±0.5 dB in the range 0° to 50°C.

5.2 Input and Output Match

As previously discussed, to prevent loss of power due to impedance mismatches the input and output impedance of a device should be equal to the characteristic impedance (e.g., 50Ω) of the source and load to which it is connected. For example, a feedforward amplifier is usually placed between a

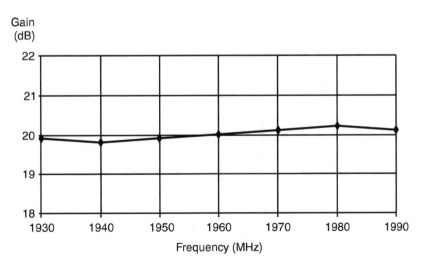

Figure 5.3 Gain versus frequency at 1-W output power (courtesy of Telia SA).

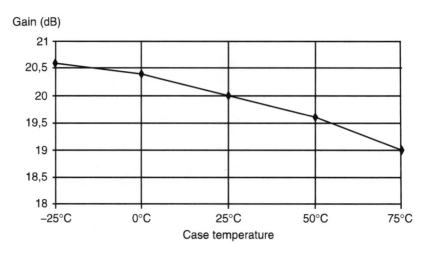

Figure 5.4 Gain versus temperature at 10-W output power (courtesy of Telia SA).

transceiver unit (the source) and a duplex filter (the load) with the actual physical connection done by coaxial cables (usually having 50Ω characteristic impedance).

A perfect impedance match at the input and output of a feedforward amplifier would result in zero power loss; however, in practice, a portion of the incident signal is reflected, resulting in a loss of transmission power. For example, Figure 5.5 shows an example of the input and output return loss of

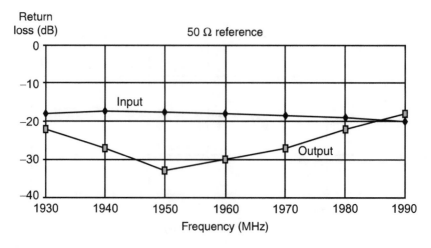

Figure 5.5 Return loss versus frequency (courtesy of Telia SA).

a feedforward amplifier; the input return loss is around −18 dB and the output return loss is better than −20 dB over the frequency band of interest.

In addition to loss of transmission power, impedance mismatches also cause frequency ripple due to the formation of standing waves (the summation of incident and reflected voltages). As explained in Section 5.5, frequency ripple is one of the principal factors that limit broadband signal cancellation and hence feedforward performance.

5.2.1 Feedforward Input Match

As shown in Figure 5.6, a signal incident at the feedforward input is split into two paths: one going to a delay line and the other to the main amplifier. The delay line signal is attenuated by the insertion loss of the coupler A_3, the loss of the delay element L_1, and the insertion loss of the coupler A_4. When Loop 1 is balanced, however, this signal is canceled at the output of coupler C_4 before it reaches the error amplifier and there is no reflected signal (irrespective of Γ_{error}).

For good noise performance (Section 5.4), the input coupler C_3 is normally chosen to have minimum loss in the delay line path. Signals going to the main amplifier are thus attenuated such that the input match Γ_{main_in} can be ignored. For example, if $C_3 = -10$ dB and the input return loss of the main amplifier is −15 dB, then the reflected signal at the feedforward

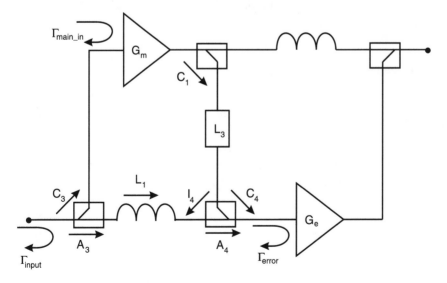

Figure 5.6 Feedforward input match.

input is attenuated by 35 dB relative to the input signal and can be considered negligible.

Theoretically, none of the signals passing through the main amplifier should reappear at the feedforward input either; however, in practice, coupler C_4 has a finite directivity D_4 (isolation I_4). That is, a small portion of the signal from the main amplifier appears at the input to coupler C_4 and travels back to the feedforward input attenuated by L_1 and A_3. This reflected signal is at a lower level than the incident signal by an amount equal to the coupler directivity (the difference between the coupler isolation and the coupling factor) plus insertion loss.

The total attenuation of a signal arriving back at the feedforward input, that is, the return loss, is therefore equal to

$$RL_{\text{input}} = 2 \cdot A_3 + 2 \cdot L_1 + A_4 + D_4 \qquad (5.11)$$

For a 1-dB increase in delay line loss, the return loss improves by 2 dB; however, the noise performance of the feedforward system is degraded—that is, the noise figure increases by 1 dB (see Section 5.4). Thus, to increase the input return loss of a feedforward amplifier, it is preferable to increase the directivity of the coupler rather than increase the delay loss. Another reason for improving coupler directivity is to reduce amplitude frequency ripple since this has a negative impact on broadband signal cancellation. When signal cancellation requires that amplitudes be matched to within a fraction of a decibel, any amplitude ripple is clearly undesirable.

5.2.2 Feedforward Output Match

As shown in Figure 5.7, a signal incident at the feedforward output is attenuated by the coupler insertion loss α_2, the loss of the delay element l_2, and the coupler insertion loss a_1 before arriving at the main amplifier output. When the second feedforward loop is balanced, however, any signal reflected from the main amplifier (reflection coefficient $\Gamma_{\text{main\_out}}$) is canceled at the feedforward output in the same way as any intermodulation and noise.

In practice, coupler C_1 has a finite isolation I_1 and therefore some signal appears at the coupled port and continues through the error amplifier path to reappear at the feedforward output. The output return loss is then equal to

$$RL_{\text{output}} = \alpha_2 \cdot l_2 \cdot i_1 \cdot l_3 \cdot c_4 \cdot g_e \cdot c_2 \qquad (5.12)$$

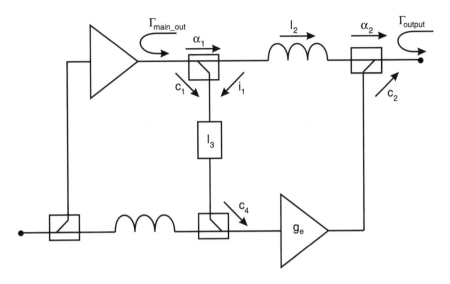

Figure 5.7 Feedforward output match.

Substituting for g_e when Loop 2 is balanced and $I_1 = d_1 c_1$ gives

$$RL_{\text{output}} = \alpha_1 \cdot \alpha_2^{\,2} \cdot l_2^{\,2} \cdot d_1 \tag{5.13}$$

Expressed in decibels

$$RL_{\text{output}}(\text{dB}) = 2 \cdot A_2 + 2 \cdot L_2 + A_1 + D_1 \tag{5.14}$$

Increasing the delay line loss improves the return loss; however, the output power of the main amplifier must then be increased to compensate and the system efficiency is then reduced. As previously explained in regard to feedforward input match, it is preferable to increase the directivity of the coupler to improve return loss and reduce amplitude frequency ripple for better broadband signal cancellation rather than increase delay loss.

5.3 Reverse Intermodulation

Reverse intermodulation is the name given to describe the intermodulation between the normal output signal of an amplifier, single or multicarrier, and an unwanted or parasitic signal incident at the output.

In a feedforward amplifier the parasitic signal travels back to the output of the main amplifier and generates intermodulation products in addition to those already present due to nonlinearities in the main amplifier. Any reverse intermodulation from the main amplifier is canceled in the second feedforward loop in exactly the same way as intermodulation generated by the main amplifier. The parasitic signal is itself attenuated by reflection at the output of the main amplifier (by an amount equal to the return loss) and additionally by second loop cancellation. A circulator can be placed at the output of the feedforward amplifier; however, this introduces additional loss and care must be taken to ensure that no distortion is added to the output signal.

For test purposes, the parasitic signal is usually an unmodulated CW carrier at a frequency offset slightly from the carrier frequencies; the typical level of the parasitic signal is around 30 dB below the carriers. Note that generation of such a signal may require another high-power amplifier since the parasitic signal cannot be connected directly to the feedforward output rather via an additional output coupler and circulator.

5.4 Noise Figure

The signal-to-noise ratio at the input to a feedforward amplifier can be defined as

$$SNR_{input} = \frac{S}{n_i} \qquad (5.15)$$

where

S = feedforward input signal power (W)
n_i = feedforward input noise power (W)

The input noise power n_i represents the maximum available noise power for a conjugate match. That is,

$$n_i = k \cdot T \cdot B \qquad (5.16)$$

where

k = Boltzmann's constant, 1.38×10^{-23} (J/K)
T = absolute temperature (K)
B = equivalent noise bandwith (Hz)

Additional noise, which is generated in the main amplifier path (noise factor f_m) by the coupler C_3 and main amplifier, behaves as an unwanted signal and is canceled when Loop 2 is balanced, that is,

$$n_i \cdot g_m \cdot f_m \cdot \alpha_1 \cdot l_2 \cdot \alpha_2 - n_i \cdot g_m \cdot f_m \cdot c_1 \cdot l_3 \cdot c_4 \cdot g_e \cdot c_2 = 0 \quad (5.17)$$

Additional noise produced in the error amplifier path (noise factor f_e) cannot, however, be canceled since there is no equivalent cancellation signal. The total noise power at the output of the error amplifier is given by

$$n_e = n_i \cdot g_e \cdot f_e \quad (5.18)$$

Rewriting (5.18) illustrates that the output noise power effectively consists of two components, that is,

$$n_e = n_i \cdot g_e + \left(f_e - 1\right) n_i \cdot g_e \quad (5.19)$$

The first term, $n_i g_e$, is due to the noise at the feedforward input and cannot be reduced (for a given temperature T and bandwidth B) since n_i effectively represents true thermal noise. The second term, $n_i g_e (f_e - 1)$, represents unwanted additional noise and can be reduced by minimizing the noise factor f_e of the error amplifier path (in the limit, $f_e \rightarrow 1$).

The SNR at the feedforward output is therefore

$$SNR_{output} = \frac{g_{ff} S}{n_e \cdot c_2} \quad (5.20)$$

Substituting for g_{ff} in (5.20)—that is, if the second feedforward loop is balanced—gives

$$SNR_{output} = \frac{\alpha_3 \cdot \alpha_4 \cdot l_1}{n_i \cdot f_e} \cdot S \quad (5.21)$$

The noise factor of the feedforward amplifier is given by the ratio of input and output SNRs, that is,

$$f_{\text{ff}} = \frac{SNR_{\text{input}}}{SNR_{\text{output}}} = \frac{f_e}{\alpha_3 \cdot \alpha_4 \cdot l_1} \qquad (5.22)$$

The noise figure F (dB) is equal to $10 \log(f)$ and (5.22) then becomes

$$F_{\text{ff}}(\text{dB}) = F_e - A_3 - A_4 - L_1 \qquad (5.23)$$

That is, the feedforward noise figure is equal to the loss from the feed-forward input through to the error amplifier plus the noise figure of the error amplifier itself. For example, if the coupler insertion losses and delay line loss are as shown in Figure 5.2 (0.5 dB, 3 dB, and 0.5 dB, respectively) and the error amplifier has a noise figure of 5 dB, then the overall feedforward noise figure is 9 dB. Increasing the loss of the delay line, for example, by using a smaller diameter cable, perhaps for mechanical reasons, increases the noise factor by 1 dB for each decibel of additional loss. As mentioned in Chapter 4, the potentially very low-noise figure of feedforward amplifiers makes them attractive for use in cable systems and wideband receivers.

5.5 Broadband Signal Cancellation

In each feedforward loop, there are two signal paths; a reference path containing only passive elements (couplers and delay lines), and an active path that includes an amplifier and gain and phase adjustment networks. Ideally, signals would be canceled by an infinite amount over an unlimited bandwidth; however, the practical difficulties of accurately matching phase, amplitude, and delay impose limits. That is, in each feedforward loop, signals can only be canceled (suppressed) by a finite amount over a finite bandwidth.

Ideal broadband signal cancellation implies that at all frequencies of interest, the reference and active path signals have:

- Equal amplitude;
- Opposite phase;
- Equal delay.

The requirements on amplitude and phase matching at a single frequency have already been discussed in Chapter 4, however the concept of equal delay requires further explanation.

5.5.1 Delay

As previously stated, the delay of a network is equal to the rate of change of signal phase with respect to frequency, that is,

$$\tau = \frac{\Delta\phi}{\Delta\omega} \tag{5.24}$$

A single-frequency signal, for example an unmodulated carrier with frequency f_0, has a delay, τ_0, equal to

$$\tau_0 = \left(\frac{\Delta\phi}{\Delta\omega}\right)_{f_0} \tag{5.25}$$

Information-bearing signals, such as multicarrier signals or modulated carriers, consist of a group of frequencies and (5.24) then refer to the group delay. Note that two signals have equal delay when their phase-frequency slopes are equal and they have constant delay when the slope is constant.

As previously mentioned in Chapter 4, the delay of an amplifier has two component parts, that is the actual transmissive delay (or baseband phase shift) and the phase shift introduced by poles and zeros in the transfer function (RF phase rotation). The delay of a network therefore increases as the number of poles increases or the transmissive delay increases.

In a feedforward amplifier, delay elements placed in the reference paths are used to compensate for the group delay through the active paths (i.e., the amplifiers and gain/phase adjustment networks). The delay elements are lengthened or shortened until a delay match is achieved, that is, until the slopes of the phase-frequency response are equal through the two paths. When the signal paths are matched with this unique value of delay, the signal cancellation is independent of frequency, as shown in the example given in Figure 5.8. If the delay through the reference path is too long or too short there is a delay mismatch and the cancellation becomes more narrowband. For example, Figure 5.9 shows the band edge suppression (i.e., suppression at 430 MHz) as a function of delay mismatch for the case given in Figure 5.8.

Note, however, that even with a delay mismatch present, it is still possible to achieve high levels of suppression over a narrow bandwidth. In certain cases, delay mismatch can even be introduced deliberately, for example to reduce power dissipation in the delay elements and hence improve overall feedforward efficiency (the penalty is more narrowband cancellation).

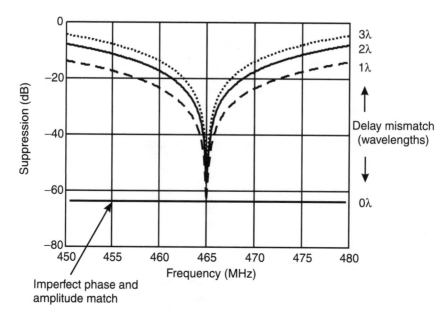

Figure 5.8 Suppression and delay mismatch.

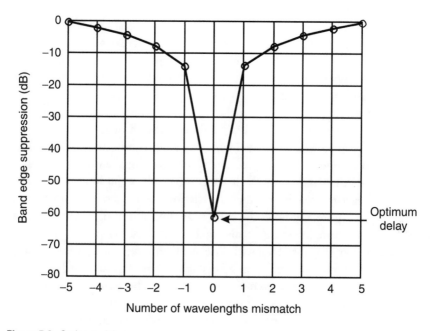

Figure 5.9 Optimum delay.

5.5.2 Frequency Response Characteristics

In terms of frequency response characteristics, broadband signal cancellation requires that at each cancellation frequency the signals have equal gain and opposite phase. The frequency dependence of the two signal paths is therefore unimportant as long as they track each other. An ideal frequency characteristic, however, would have the following characteristics:

- Constant gain over the cancellation bandwidth;
- Linear phase over the cancellation bandwidth.

Note that a network that has a linear phase response has a constant delay; the phase-frequency response is a straight line with a constant slope equal to the delay.

A low-pass network exhibits constant gain and linear phase at frequencies well below cut-off. For example, Figure 5.10 shows the amplitude and phase response of a low-pass network having a transfer function of the form given in (5.26); for simplicity the delay τ is assumed zero and $n = 1$, that is, a single pole,

$$T(\omega) = \frac{1}{\left(1 + j \cdot \dfrac{\omega}{\omega_L}\right)^n} \cdot e^{-j \cdot \omega \cdot \tau} \tag{5.26}$$

As Figure 5.10 shows, the phase is linear, that is, the delay is constant only at frequencies well below cut-off (usually much higher than the RF frequencies of interest). As the number of poles, n, or the transmissive delay, τ, increases, the delay of the network is increased and the phase changes more rapidly. Each pole contributes to the overall response with an amplitude roll-off factor of 20 dB/decade and a phase shift of 90 degrees; typical transmissive delays for high-power RF amplifiers are in the nanosecond range.

Due to their constant gain and linear phase characteristics, low-pass structures are ideally suited as delay elements in feedforward amplifiers. Increasing or decreasing the amount of delay simply changes the slope of the phase-frequency response but in each case the phase remains linear and the delay remains constant. Examples of low-pass structures commonly used for feedforward delay matching include coaxial cables, microstrip line, and stripline.

The active paths of a feedforward amplifier include amplifiers which are characterized by a band-pass rather than low-pass frequency response. Figure

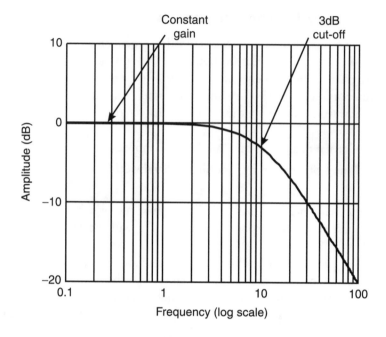

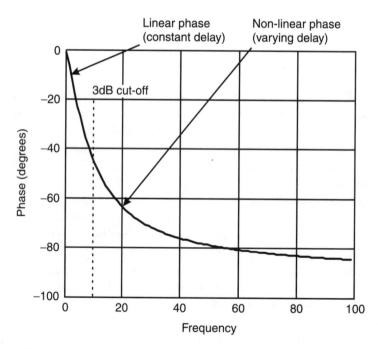

Figure 5.10 Low-pass network frequency response (single pole).

5.11 shows the amplitude and phase responses of a two-pole band-pass network ($\tau = 0$), that is,

$$T(\omega) \frac{1}{\left(1 + j \cdot \dfrac{\omega}{\omega_L}\right) \cdot \left(1 - j \cdot \dfrac{\omega_H}{\omega}\right)} \cdot e^{-j \cdot \omega \cdot \tau} \qquad (5.27)$$

When plotted on a logarithmic scale the amplitude response is symmetrical on a center frequency equal to the geometric frequency, that is,

$$\omega_c = \sqrt{\omega_H \cdot \omega_L} \qquad (5.28)$$

The delay is constant only well within the passband; deviation from linear phase (i.e., nonconstant delay) occurs as the amplitude rolls off at the band edges. An important consequence of phase rotation and amplitude roll-off is that in order to ensure constant delay, the passband of an amplifier in a feedforward system should be much wider than the cancellation bandwidth. For example, at 2 GHz it is typically necessary for an amplifier to have constant gain over ≈ 50 MHz on either side of the cancellation bandwidth.

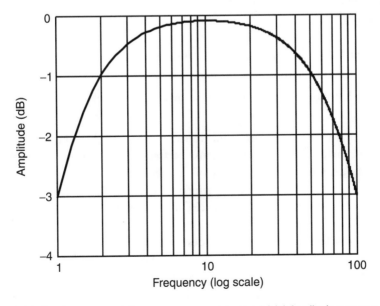

Figure 5.11 Band-pass network frequency response (single pole) (a) Amplitude response.

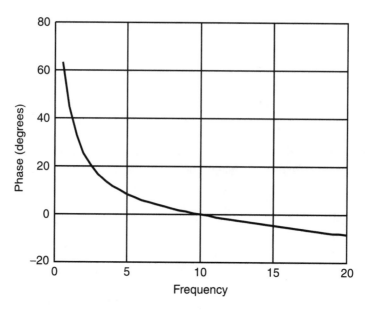

Figure 5.11 (continued) (b) Phase response.

5.5.3 Amplitude Flatness

Amplitude flatness is a measure of deviation from constant gain and is affected by:

- Amplitude ripple across the bandwidth of interest;
- Amplitude roll-off at the band edges;
- Amplitude slope across the bandwidth of interest.

Standing waves arising from impedance mismatches and/or imperfect device behavior are the primary cause of frequency ripple while amplitude roll-off is determined by the delay of the network or, more specifically, the number of network poles. As explained in Chapter 1, the frequency slope arises because the gain of transistor circuits, for example amplifiers, typically drops with frequency.

In any feedforward loop, the desired loop suppression determines the required amplitude flatness. For example, the amplitude flatness must be better than 0.5 dB to achieve 25-dB cancellation and 0.1 dB for 40-dB cancellation. The practically achievable amplitude flatness of the components in a feedforward path may be improved by using a compensation network. The compensation network ensures that the overall amplitude slope of the feedforward path is within the specified limit by introducing a slope with the opposite gradient. Note, however, that compensation networks introduce extra delay and thus increase the loss in the corresponding delay element.

5.5.4 Phase Flatness

Phase flatness is a measure of deviation from linear phase and hence deviation from constant delay. As shown in Figure 5.12, the actual phase is determined by evaluating the argument of the network transfer function $T(\omega)$; the linear phase is calculated assuming that the delay is constant and equal to the delay at some chosen reference frequency ω_0. That is, phase flatness $\phi(\omega)$ is given by

$$\phi(\omega) = \arg(T(\omega)) - \tau_0 \cdot \omega \qquad (5.29)$$

The delay τ_0 is effectively an electrical length compensation term and is equal to the delay at the reference frequency ω_0, that is,

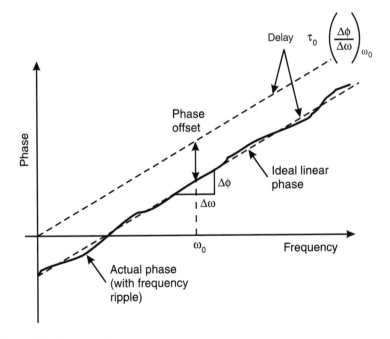

Figure 5.12 Phase flatness.

$$\tau_0 = \left(\frac{\Delta\phi}{\Delta\omega} \right)_{\omega_0} \qquad (5.30)$$

In practice, the electrical length compensation function on a network analyzer can be used to display phase flatness, and the difference between the actual phase and the linear phase at the reference frequency is called the phase-offset. The phase offset is equal to zero for a low-pass network such as a cable; but in principle, it can take any value depending upon the exact form of the transfer function.

5.5.5 Group Delay, Group Velocity, and Dispersion

Multicarrier or modulated signals consist of a group of frequencies and form a wave packet described by the envelope voltage; a two-tone signal, as described in Chapter 2, is a simple example. The wave inside the envelope propagates with a velocity equal to the *phase velocity*, v_p, while the velocity of the envelope is equal to the *group velocity*, v_g. That is,

$$v_g = \frac{1}{\left(\dfrac{\Delta\beta}{\Delta\omega} \right)} \qquad (5.31)$$

$$v_p = \frac{\omega_0}{\beta_0} \qquad (5.32)$$

Similarly, the group delay, τ_g, represents the delay of the signal envelope and τ_0 represents the delay of the wave inside the envelope (corresponds to the phase velocity). That is,

$$\tau_g = \frac{\Delta\phi}{\Delta\omega} \qquad (5.33)$$

$$\tau_0 = \left(\frac{\Delta\phi}{\Delta\omega} \right)_{\omega_0} \qquad (5.34)$$

The phase constant β (units rad/m) in (5.31) plays a similar role to the phase term f in (5.33). When the *phase constant* β is a linear function of frequency, all frequency components travel at the same velocity; the phase velocity is equal to the group velocity and there is no distortion. When the phase constant is not a linear function of frequency, different frequency components travel at different velocities; the phase velocity is not equal to the group velocity ($u_g < u_p$), and there is "dispersive" distortion.

Similarly, considering delay rather than velocity, if the phase of a network is linear, then all frequency components have the same delay; the individual delays are equal to the group delay and there is no distortion. If, however, the phase is nonlinear, then different frequency components are delayed by different amounts; the group delay is not equal to the delay of individual frequency components and there is dispersive distortion. Dispersive distortion can thus be thought of in terms of either nonconstant velocity or nonconstant delay; in both cases, the important parameter is the linearity of the phase-frequency relationship.

In the context of RF circuits, lossy dielectrics, transmission lines, and band-pass filters are all examples of dispersive mediums. For feedforward amplifiers in particular, the resulting phase ripple inevitably causes some reduction in broadband signal cancellation and an increase in output signal distortion.

5.5.6 Broadband Cancellation Examples

Signal cancellation involves vector addition and therefore the resultant, or suppressed signal, is also a vector. The amplitude term is usually of most practical interest since it determines the power level of the suppressed signal, for example, the residual carrier power in the error amplifier path and the distortion power level at the feedforward output. The phase term can also provide very useful information since delay mismatch and amplitude and phase flatness affect the suppressed signal in different ways. That is, the polar response (amplitude and phase combined) of the suppressed signal gives more insight into the practical causes of limited broadband suppression than the magnitude response alone.

Table 5.1 shows different combinations of three parameters that affect broadband signal cancellation, namely:

- Amplitude ripple (nonconstant gain);
- Phase ripple (variable gradient, nonconstant delay);
- Delay mismatch (mismatch in slopes of the phase-frequency response).

Table 5.1
Broadband Cancellation Examples

Example	Amplitude mismatch	Phase mismatch	Delay mismatch	Figure
1	0.5 dB	0	0	5.13
2	0	3.5 degrees	0	5.13
3	0	0	1λ	5.14
4	1 dB	8 degrees	1λ	5.15

In each case, perfect amplitude and phase matching is assumed at the center frequency (in practice provided by the gain and phase adjustment networks). In these examples, the center frequency is 465 MHz and the cancellation bandwidth is 30 MHz, but the same principles apply at other frequencies and for other cancellation bandwidths.

In Figure 5.13(a), which shows the signal cancellation versus frequency response, although only one curve is visible, there are actually two curves indistinguishable from each other (Examples 1 and 2 in Table 5.1). The ampli-

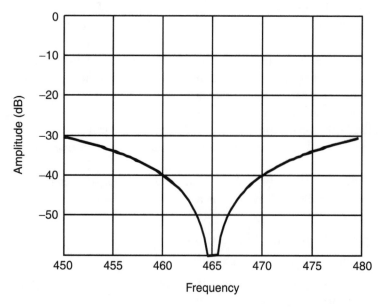

Figure 5.13 Broadband suppression with phase and amplitude ripple (a) Amplitude response.

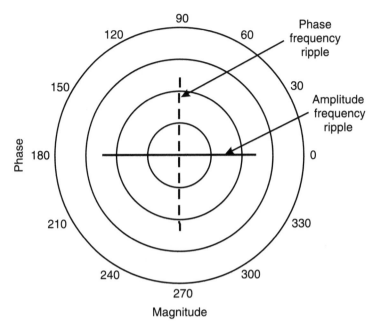

Figure 5.13 (continued) (b) Polar response.

tude responses in both cases are identical, that is, 30-dB cancellation at the band edges (30-MHz separation) and 40-dB cancellation over the center 10 MHz. If, however, the phase response is considered in addition to the amplitude response, then it is possible to differentiate between the two examples; see Figure 5.13(b), which shows the polar response. Note that the polar representation of a signal shows both the amplitude and phase—the loci indicate the variation over frequency.

Example 1, which represents signal cancellation involving ideal phase and delay characteristics but nonideal amplitude, is a straight line on the real axis. Example 2, which represents signal cancellation involving ideal amplitude and delay characteristics but nonideal phase, is a straight line on the imaginary axis. That is, the real axis of the polar response is associated with nonideal broadband amplitude characteristics, while the imaginary axis represents nonideal broadband phase characteristics.

Figure 5.14 show the amplitude and polar response for Example 3 in Table 5.1. That is, signal cancellation with perfect broadband, amplitude, and phase characteristics but a delay mismatch of one wavelength at the center frequency. The polar response for Example 3 (nonideal delay) is confined to the imaginary axis in a similar manner to that shown in Example 1

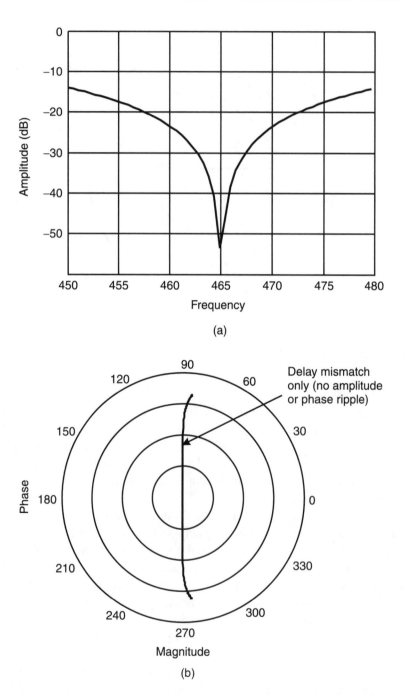

Figure 5.14 Broadband cancellation with delay mismatch (a) Amplitude response (b) Polar response.

(nonideal phase). In both cases however, as the amount of mismatch increases (or the cancellation bandwidth), the polar response has a progressively larger real part; that is, the response is no longer a straight line on the imaginary axis. As previously mentioned, delay is simply the rate of change of phase and therefore it is not surprising that both parameters affect the polar response in a similar manner.

For illustrative purposes, the previous examples have assumed ideal broadband signal characteristics for two of the three parameters—amplitude, phase, and delay; however, in practice, all exhibit some nonideal behavior. For example, Figure 5.15 shows a general case (Example 4 in Table 5.1) and it can be seen that the polar response curve has both real and imaginary parts. The real part is due to amplitude mismatch, that is, amplitude ripple; and the imaginary part is the result of imperfect delay matching and phase ripple.

From a practical viewpoint, the polar response is very useful in determining the causes of nonideal broadband suppression, that is, whether the amplitude or phase response needs to be improved, or the delay match, or a combination. Furthermore, the polar response also provides a method of adjusting or trimming the delay to compensate for variations between feedforward amplifiers in a production environment.

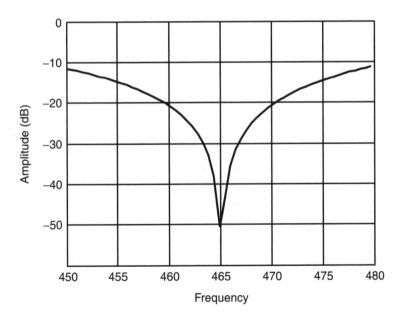

Figure 5.15 Broadband cancellation—general case (a) Amplitude response.

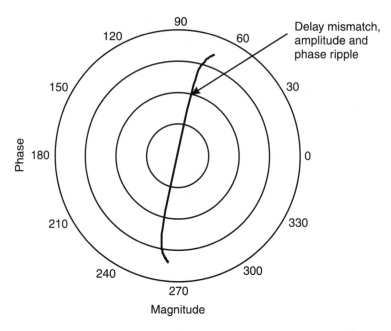

Figure 5.15 (continued) (b) Polar response.

5.5.7 Delay Matching—Frequency Marker Example

A network analyzer can be used to display the amplitude and polar responses as shown in the previous examples. Moreover, in using frequency markers it is possible to determine if the delay is due to a positive or negative delay mismatch, that is, if the delay element is too long or too short. Using the "marker delta function" on a network analyzer, it is also possible to quantify the delay mismatch in terms of, for example, the number of nanoseconds mismatch.

For example, Figure 5.16 shows the broadband polar response with frequency markers placed at the lower and upper ends of the cancellation band (450 MHz and 480 MHz, respectively). The position of the frequency markers depends upon whether the delay mismatch is positive or negative, and the marker delta function on the network analyzer can be used in practice. Figure 5.16 also shows the polar response when the delay mismatch is zero (assuming no phase and amplitude mismatch). Delay trimming can be done using the frequency markers in real and imaginary mode and calibrating, that is, a certain number of milliunits on the imaginary scale corresponds to a certain delay (an electrical length, e.g., a length of cable).

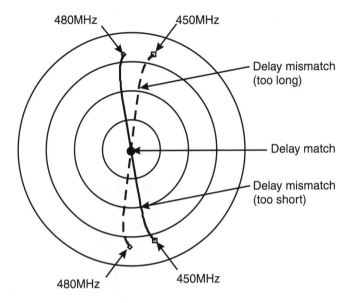

Figure 5.16 Broadband cancellation—frequency marker example.

To summarize, delay mismatch and frequency-dependent effects such as amplitude and phase ripple prevent signals from being precisely matched over a wide bandwidth and ultimately set the practical limit on the otherwise theoretically unrestricted bandwidth of feedforward operation. The wider the bandwidth, the harder it is in practice to reduce amplitude and phase ripple to within the fraction of a decibel and a degree necessary to achieve high values of loop suppression.

5.6 Error Amplifier Power

The average and peak-power requirement for the main amplifier in a feedforward system has already been discussed, however, the question arises as to the equivalent requirements on the error amplifier.

5.6.1 Average Power Requirement

In a feedforward system, the input signal to the error amplifier consists of intermodulation and noise from the main amplifier and residual carrier power from the carrier suppression loop (Loop 1). Depending upon the nature of the loop control scheme, there may also be a *pilot signal* (Section 5.9).

The output distortion power, P_D, of a typical Class AB amplifier (Figure 5.17) with an average output power, P_M, and distortion level, D_M, can be approximated by

$$P_D = P_M \cdot 3 \cdot 10^{\frac{D_M}{10}} \tag{5.35}$$

That is, the distortion power at the output of the main amplifier is approximately three times the level of the highest intermodulation product relative to the total, average output power.

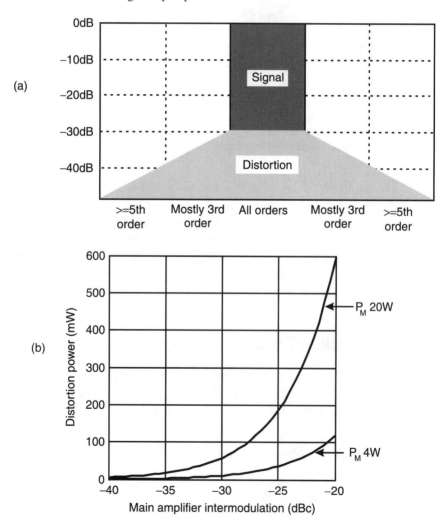

Figure 5.17 Distortion power estimation (a) Frequency distribution (b) Calculation example.

The power gain (or rather the loss) from the output of the main amplifier to the feedforward output is determined by the insertion loss of the couplers, α_1 and α_2, and the delay line, l_2, that is,

$$\alpha_1^{\,2} \cdot l_2^{\,2} \cdot \alpha_2^{\,2} \tag{5.36}$$

When the second loop is balanced, the gain through the error amplifier path is equal to the gain through the delay line path. The average output power of the error amplifier is then given by

$$P_E = P_M \cdot \left(3 \cdot 10^{\frac{D_M}{10}} + 10^{\frac{S_C}{10}} \right) \cdot \alpha_1^{\,2} \cdot l_2^{\,2} \cdot \alpha_2^{\,2} \cdot \frac{1}{c_2^2} \tag{5.37}$$

Equation (5.37) also includes a term S_c, which represents the residual carrier power from the carrier cancellation loop; the target level for carrier suppression is typically −30 dBc over the carrier bandwidth. Depending upon the relative performance of the main amplifier and the carrier loop suppression, the dominant contribution to the average output power of the error amplifier can be either distortion power or residual carrier power.

For example, consider a feedforward amplifier with the parameters shown in Table 5.2.

Evaluating (5.37), the total, average output power of the error amplifier in this example is 180 mW (44-mW residual carrier and 134-mW inter-

Table 5.2
Error Amplifier Average Power Example

Main amplifier average power	P_M	10W
Main amplifier distortion	D_M	−30 dBc
Carrier suppression	S_c	30 dB
Main amplifier coupler		
Coupling factor	C_1	−20 dB
Insertion loss	A_1	−0.05 dB
Output coupler		
Coupling factor	C_2	−10 dB
Insertion loss	A_2	−0.5 dB
Delay line loss	L_2	−3 dB

modulation distortion). If the loop suppression drops to 25 dBc, then the total power rises to 275 mW, of which 231 mW is residual carrier power.

5.6.2 Peak-Power Requirement

The peak-power requirement of the error amplifier is higher than the average power by an amount equal to the peak-to-average ratio of the main amplifier ΔP_M, and an additional margin, B_E, to ensure linear operation, that is,

$$P_{1E} = P_E \cdot B_E \cdot \Delta P_M \qquad (5.38)$$

The back-off factor B_E, typically 6 dB, is necessary to ensure that the error amplifier contributes as little as possible to the intermodulation at the feedforward output. Any intermodulation (or noise) created by the error amplifier, or indeed any leakage signals picked up by the error amplifier, go straight to the feedforward output with no possibility of cancellation. The intermodulation level of the error amplifier depends upon the target system linearity and the coupling factor C_2. In practice, intermodulation levels of ≈ -60 dBc at average output power may be required.

To minimize distortion due to signal clipping in the main amplifier, the peak-to-average ratio of the main amplifier is designed to be equal to the peak-to-average ratio of the feedforward input signal, ΔP_S. That is,

$$\Delta P_M = \frac{P_{1M}}{P_M} = \Delta P_S \qquad (5.39)$$

Substituting for ΔP_M, the peak power of the error amplifier can then be expressed in terms of the peak power of the main amplifier, that is,

$$P_{1E} = P_{1M} \cdot B_E \cdot \left(3 \cdot 10^{\frac{D_M}{10}} + 10^{\frac{S_C}{10}} \right) \cdot \alpha_1{}^2 \cdot l_2{}^2 \cdot \alpha_2{}^2 \cdot \frac{1}{c_2{}^2} \qquad (5.40)$$

For feedforward systems that use a pilot signal, (5.40) is modified to become

$$P_{1E} = P_{1M} \cdot B_E \cdot \left(3 \cdot 10^{\frac{D_M}{10}} + 10^{\frac{S_C}{10}} + 10^{\frac{P_L}{10}} \right) \cdot \alpha_1{}^2 \cdot l_2{}^2 \cdot \alpha_2{}^2 \cdot \frac{1}{c_2{}^2} \qquad (5.41)$$

P_L is the level of the pilot signal relative to the carriers.

Equations (5.40) and (5.41) state that the peak power of the error amplifier is a function of the peak-power and intermodulation performance of the main amplifier, the level of carrier suppression, and coupler and delay line losses. For example, consider a feedforward amplifier with the parameters shown in Table 5.3.

Figure 5.18 shows the effect of the intermodulation performance of the main amplifier on the peak-power requirements of the error amplifier for different loop suppressions. As the intermodulation performance of the main amplifier improves, the peak-power requirements on the error amplifier are reduced. A similar effect occurs as the carrier loop suppression increases.

Note that the relationship is nonlinear; that is, as the intermodulation performance drops below ≈30 dBc, the required peak power rises almost exponentially. Alternatively, below ≈35 dBc the intermodulation power is no longer the dominant factor in determining the peak power of the error amplifier; the residual carrier power dominates.

5.7 Error Amplifier Gain-Correction Capability

In the second feedforward loop, distortion from the main amplifier is canceled at the output of the coupler C_2 by vector cancellation; that is, distortion from the main and error amplifier paths has equal amplitude, opposite phase,

Table 5.3
Error Amplifier Peak Power Example

Main amplifier average power	P_M	10W
Main amplifier peak-to-average ratio	ΔP_M	10 dB
Main amplifier peak power	P_{1M}	100W
Main amplifier coupler		
Coupling factor	C_1	−20 dB
Insertion loss	A_1	−0.05 dB
Output coupler		
Coupling factor	C_2	−10 dB
Insertion loss	A_2	−0.5 dB
Delay line loss	L_2	−3 dB

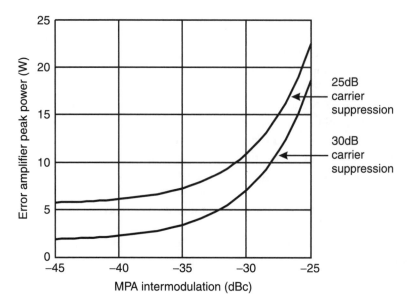

Figure 5.18 Error amplifier peak-power requirements.

and matched delay. The amount of distortion power generated by the error amplifier is a measure of its *distortion-correction* capability, however the error amplifier also has a *gain-correction* capability.

Gain-correction capability refers to the ability of the error amplifier to compensate for changes in the signal level of the main amplifier path. The power supplied to the isolated port of the output coupler, C_2 is used to "modulate" the feedforward output signal and hence provide constant feedforward gain.

5.7.1 Voltage and Power Distribution

Figure 5.19 shows the voltage distribution at the output coupler; the two input voltages, V_{mpa} and V_{epa}, come from the main and error amplifiers, respectively, and can contain both carrier and intermodulation signals. The two output voltages, V_{out} and V_{iso}, are the feedforward output voltage and the voltage at the isolated port

$$V_{out} = c \cdot V_{epa} + j \cdot \alpha \cdot V_{mpa} \tag{5.42}$$

$$V_{iso} = c \cdot V_{mpa} + j \cdot \alpha \cdot V_{epa} \tag{5.43}$$

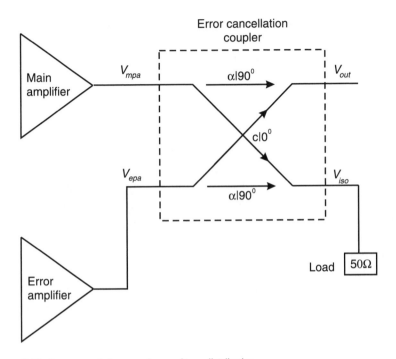

Figure 5.19 Error cancellation coupler—voltage distribution.

Consider first the case when V_{mpa} contains only carrier signals. When the first feedforward loop is balanced, the carrier output of the error amplifier is zero ($V_{epa} = 0$) and the signal appearing at the feedforward output and the isolated port of the coupler are equal to

$$V_{out} = j \cdot \alpha \cdot V_{mpa} \tag{5.44}$$

$$V_{iso} = c \cdot V_{mpa} \tag{5.45}$$

For example, if the coupling coefficient is 10 dB, then 90% of the carrier power appears at the feedforward output and 10% goes to the isolated port and becomes heat.

Now consider the case when V_{mpa} and V_{epa} consist of only intermodulation. If the reference voltage, V_{mpa}, is defined as $V_{mpa} = j$, and V_{epa} is a real quantity, the phase difference at the coupler output between the main and error amplifier paths is 180 degrees, the condition for signal cancellation. Thus, assuming equal amplitude (e.g., 3-dB coupling coefficient and

$V_{epa} = 1$), any intermodulation power from the error amplifier is effectively canceled at the output port due to the 180-degree phase relationship.

Now consider the case when the gain of the main amplifier and hence its output power changes for some reason, that is,

$$V_{mpa} = (1 - \delta) \cdot V_{mpa} \qquad (5.46)$$

The gain of the main amplifier may change for a number of reasons including nonlinear and temperature effects; for example, as previously discussed, a Class AB amplifier has a variable gain depending on signal level. Unless a dynamic loop control scheme is used (Chapter 6), the first feedforward loop is then only matched for the nominal gain of the amplifier and hence, the overall feedforward gain varies with signal level. In order to maintain constant feedforward output power and constant feedforward gain, the error amplifier has to supply the necessary power (in-phase or out-of-phase depending upon whether the main amplifier gain increases or decreases).

When the main amplifier signal compresses, the output voltage becomes (substituting (5.46) in (5.42))

$$V_{out} = c \cdot V_{epa} + j \cdot \alpha \cdot V_{mpa} \cdot (1 - \delta) \qquad (5.47)$$

Setting V_{out} in (5.47) equal to V_{out} in (5.44), that is, maintaining a constant output power and hence constant gain, the error amplifier voltage V_{epa} becomes

$$V_{epa} = j \cdot V_{mpa} \cdot \frac{\delta \cdot \alpha}{c} = j \cdot V_{mpa} \cdot \delta \cdot \frac{\sqrt{1 - c^2}}{c} \qquad (5.48)$$

Now substituting V_{epa} back into (5.47), the output voltage becomes

$$V_{out} = j \cdot c \cdot V_{mpa} \cdot \delta \cdot \frac{\sqrt{1 - c^2}}{c} + j \cdot \alpha \cdot V_{mpa} \cdot (1 - \delta)$$
$$V_{out} = j \cdot V_{mpa} \cdot \sqrt{1 - c^2} \qquad (5.49)$$

That is, the output voltage is *constant* and *independent* of the change in main amplifier gain.

At the isolated port the voltage becomes

$$V_{iso} = c \cdot V_{mpa} \cdot (1 - \delta) + j \cdot \alpha \left(j \cdot V_{mpa} \cdot \frac{\delta \cdot \alpha}{c} \right)$$

$$V_{iso} = c \cdot V_{mpa} \cdot \left(1 - \frac{\delta}{c^2} \right) \tag{5.50}$$

Unlike the feedforward output voltage V_{out}, the voltage and hence the power at the isolated port (dissipated as heat) is dependent upon the degree of unbalance (δ) in the first feedforward loop. When $\delta = c^2$, the power in the isolated port and hence the heat dissipation is a minimum. For example, if the coupling coefficient $C = -10$ dB, then $c = 0.316$ and $\delta = c^2 = 0.1$.

Figure 5.20 shows the power distribution in an output coupler, which has a coupling factor $C = -10$ dB, as a function of compression factor δ. When the main amplifier signal compresses (i.e., the gain drops), δ is positive; if the gain increases, then δ is negative. The nominal input power from the main amplifier is $V_{mpa}^2 = 1.0$ ($|V_{mpa}| = 1$) and the nominal feedforward output power (after the insertion loss of the coupler) is $V_{out}^2 = 0.9$.

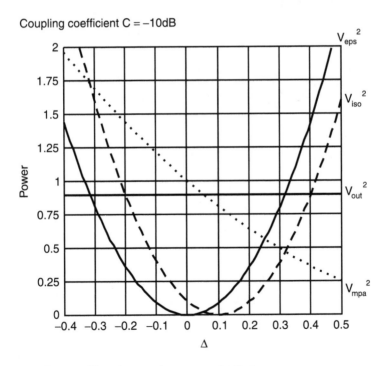

Figure 5.20 Error amplifier gain correction—power distribution.

As δ increases, $(1 - \delta)$ decreases, and the power in the isolated port drops to a minimum $(\delta = c^2)$ before rising again. The output power remains constant regardless of δ and, as δ increases further, the error amplifier supplies more power until the main amplifier is effectively redundant (i.e., $V_{mpa} \rightarrow 0$). Note that when the second loop is balanced, the gain through the error amplifier path (from feedforward input, through the delay line) is equal to the feedforward gain (Section 5.1). In principle, it is therefore possible to turn off the main amplifier at low signal levels and just use the error amplifier. Remember that the main amplifier is primarily used to produce high power. If the input signal has a high dynamic range, then it may be possible to avoid using the main amplifier at low power levels where the efficiency is low.

Figure 5.20 also shows that as δ decreases (i.e., $(1 - \delta)$ increases), the error amplifier produces more power, but this time in antiphase in order to maintain constant output power. Note also that the power in the isolated port can become quite high and therefore the load resistor(s) in the coupler must be dimensioned correctly to avoid damage due to overheating.

5.7.2 Peak Power and Gain Correction

The analysis, so far, assumes that the error amplifier is an ideal source; in practice, however, it has a finite power output capability P_{1E}, as does the main amplifier, compression point power P_{1M}. The difference in power-handling requirements between the two amplifiers (in decibels) is given by

$$\Delta P_E = P_{1M} - P_{1E} \tag{5.51}$$

From (5.48) the ratio of error and main amplifier voltages is

$$\frac{V_{epa}}{V_{mpa}} = \frac{\delta \cdot \alpha}{c} \tag{5.52}$$

Re-arranging gives

$$\delta = \frac{c}{\alpha} \cdot \frac{V_{epa}}{V_{mpa}} = \frac{c}{\alpha} \cdot 10^{\frac{-\Delta P}{20}} \tag{5.53}$$

The maximum difference, $\delta = \delta_{max}$, occurs when $\Delta P = \Delta P_E'$, that is,

$$\delta_{max} = \frac{c}{\alpha} \cdot 10^{\frac{-\Delta P_E}{20}} \tag{5.54}$$

The gain-correction capability can then be defined as

$$GCC = 20 \cdot \log\left(\frac{1}{1 - \delta_{max}}\right) \tag{5.55}$$

Figure 5.21 shows the gain-correction GCC as a function of peak power ΔP_E for different values of the coupling coefficient C (3 dB, 6 dB, and 10 dB, respectively). The optimum value of ΔP_E for a given coupler value is when no power is dissipated in the isolated port, that is, $V_{iso} = 0$ and $\delta = \delta_{opt} = c^2$. Substituting and re-arranging gives

$$\Delta P_{opt} = 10 \cdot \log\left(1 - 10^{\frac{C}{10}}\right) - C \tag{5.56}$$

In Figure 5.22, the optimum peak-power ratio ΔP_{opt} is shown as a function of coupling coefficient C. For example, $\Delta P_{opt} = -9.5$ dB for a 10-dB

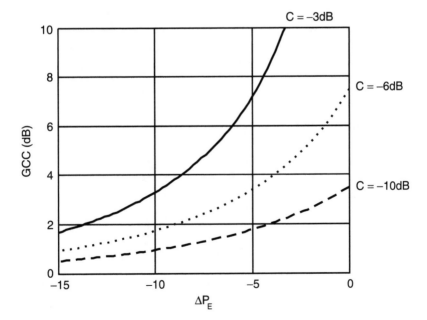

Figure 5.21 Gain correction and coupling coefficient.

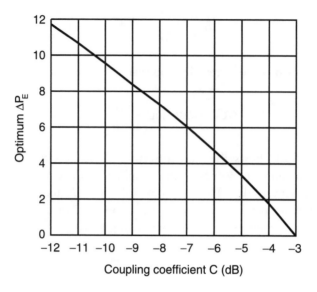

Figure 5.22 Optimum peak-power ratio.

coupler and −6 dB for a 7-dB coupler. Low values of coupling have the advantage of lower ΔP_{opt} but the disadvantage of higher insertion loss, which in turn requires the output power of the main amplifier to be increased. High values of coupling have low insertion loss but require the error amplifier to be larger relative to the main amplifier.

5.8 Feedforward Efficiency

As described in Chapter 3, the efficiency of an amplifier is a measure of how effectively dc power is converted to RF power, that is,

$$\eta = \frac{P_{out}}{P_{DC}} \tag{5.57}$$

In a feedforward amplifier there are two amplifiers, each drawing power from the dc supply; thus the total dc power is

$$P_{DC} = P_{DC\_M} + P_{DC\_E} \tag{5.58}$$

The efficiency of the main amplifier (operating in class AB) is a function of the maximum, average output power, that is,

$$P_{DC\_M} = \frac{P_M}{\eta_M} \tag{5.59}$$

The peak-to-average ratio of the main amplifier is $\Delta P_M = P_{1M} / P_M$ and the power consumption of the main amplifier can therefore be rewritten as

$$P_{DC\_M} = \frac{P_{1M}}{\eta_M \cdot \Delta P_M} \tag{5.60}$$

The efficiency of the error amplifier is

$$P_{DC\_E} = \frac{P_{1E}}{\eta_E} \tag{5.61}$$

If the ratio of error to main amplifier power capability is defined as $\Delta P_E = P_{1M} / P_{1E}$, then

$$P_{DC\_E} = \frac{P_{1M}}{\eta_E \cdot \Delta P_E} \tag{5.62}$$

Due to coupler and delay line losses, the feedforward output power P_{out} is less than the output power of the main amplifier, that is,

$$P_{out} = P_M \cdot \alpha_1^{\,2} \cdot l_2^{\,2} \cdot \alpha_2^{\,2} \tag{5.63}$$

Substituting gives the feedforward efficiency as

$$\eta_{FF} = \frac{\alpha_1^{\,2} \cdot l_2^{\,2} \cdot \alpha_2^{\,2}}{\left(\dfrac{1}{\eta_M} + \dfrac{\Delta P_M}{\eta_E \cdot \Delta P_E} \right)} \tag{5.64}$$

Feedforward efficiency is thus affected by the following factors:

- Coupler insertion losses;
- Delay line loss;
- Efficiency of the main amplifier (at maximum average output power);
- Efficiency of the error amplifier (at peak power);

- Signal peak-to-average ratio;
- Ratio of the main and error amplifier peak powers (intermodulation performance of the main amplifier and carrier suppression).

The efficiency of a practical feedforward amplifier is dependent on additional factors. For example, the power consumption of other components (digital and analog)—such as gain/phase adjustment circuits, detector circuits, low-power amplifier stages, loop control circuitry—and the efficiency of any DC/DC conversion. Furthermore, depending on whether cooling fans are an integral part of the feedforward amplifier, it may be necessary to include their power consumption in any efficiency calculation.

Figure 5.23 shows the "typical" feedforward efficiency as a function of the main amplifier efficiency—calculated using (5.64). Two different cases are considered; the first assumes a GaAs error amplifier and the second assumes a bipolar error amplifier. Table 5.4 shows the parameters used in the calculation.

In Figure 5.23, the feedforward efficiency is higher when a GaAs amplifier is used as the error amplifier rather than a bipolar amplifier (Chapter 3). The typical efficiencies of bipolar and MOSFET main amplifiers are comparable, for example 15%, and this gives an overall feedforward efficiency of

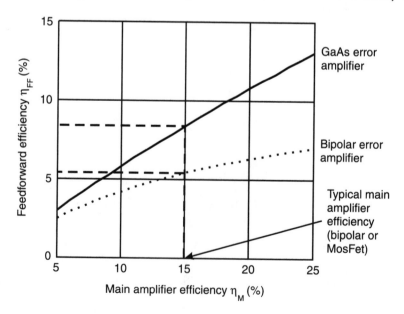

Figure 5.23 Feedforward efficiency—10-dB signal peak-to-average ratio.

Table 5.4
Feedforward Efficiency Calculation

Main amplifier peak-to-average ratio	ΔP_M	10 dB
Error amplifier peak-to-average ratio		10 dB
GaAs amplifier	ΔP_E	16 dB
Bipolar amplifier	ΔP_E	10 dB
Error amplifier efficiency		
GaAs amplifier	η_E	30%
Bipolar amplifier	η_E	20%
Main amplifier coupler		
Coupling factor	C_1	−20 dB
Insertion loss	A_1	−0.05 dB
Output coupler		
Coupling factor	C_2	−10 dB
Insertion loss	A_2	−0.5 dB
Delay line loss	L_2	−1.5 dB

8.4% and 5.4% (GaAs and bipolar error amplifiers, respectively). The efficiency of Class AB feedforward amplifiers is thus in general relatively low (5% to 10%), and therefore it is always desirable to minimize the RF output power to reduce power consumption and heat dissipation.

5.9 Loop Control

As discussed in Chapter 4, feedforward amplifiers require some form of automatic control scheme to compensate for changes in device characteristics, for example with respect to time, temperature, voltage, and signal level. Class AB amplifiers, for example, which are typically used as the main amplifier in the first feedforward loop, have a gain and phase that can vary with signal level, supply voltage, temperature, and time. Loop control is a major part of feedforward design and is one of the primary distinguishing factors between different feedforward amplifiers; the intention here is only to give a brief overview and some examples rather than a complete analysis. Note that the development and application of loop control techniques is an area where competition between different manufacturers is particularly strong (e.g., in the area of patent protection and licensing agreements).

5.9.1 General Principle

Figure 5.24 (repeated from Figure 4.13) shows the general principle of loop control whereby information on loop balance is acquired and fed back to control the loop gain and phase. Due to differences in the nature of the cancellation signal, however (in one case carriers and in the other distortion), different control techniques are often used in the first and second feedforward loops. For example, acquiring accurate information on loop balance can be very difficult in the distortion-cancellation loop and, therefore, more complicated loop control schemes involving pilot signals are often used.

In the following sections, two schemes—one analog and one digital—are presented to illustrate the control of carrier cancellation loops; the general principle of pilot signals is also explained for distortion cancellation loops. One alternative however, which does not require any direct knowledge of the loop balance, is to use look-up tables.

5.9.2 Loop Control Using Look-Up Tables

If an amplifier is sufficiently well characterized such that the values of amplitude and phase, which give good loop suppression, are tabulated as functions of certain control parameters, then it is a relatively simple task to implement loop control. For example, the temperature, input signal level, or the supply

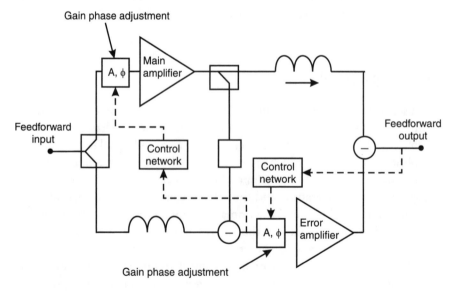

Figure 5.24 Gain/phase adjustment and loop control.

voltage can be used as control parameters. Detectors, such as a temperature sensor, are used to determine the current conditions (i.e., the current temperature), and the corresponding amplitude and phase data are then read from a look-up table and applied to the gain and phase control networks. Data collected from production calibrations could be stored in an EEPROM and the sensors checked at regular intervals to see if new control signals for the amplitude and phase networks are required.

Look-up tables have the advantage that no knowledge of the loop balance is required. The major disadvantage of look-up tables is the time it takes to fully characterize even a single amplifier; for volume production, characterizing many amplifiers, for example just over temperature, is a very time-consuming and therefore expensive process. Look-up tables do, however, provide a relatively simple and potentially very effective method of loop control, which can be applied to both carrier and distortion-cancellation loops.

5.9.3 Loop 1 Analog Control Example

Figure 5.25 shows an example of an analog control scheme for a carrier-cancellation loop whereby a sample of the reference signal (containing only carriers) is correlated with the error signal (containing distortion plus carriers). The alignment information is fed back via loop filters to the gain and phase

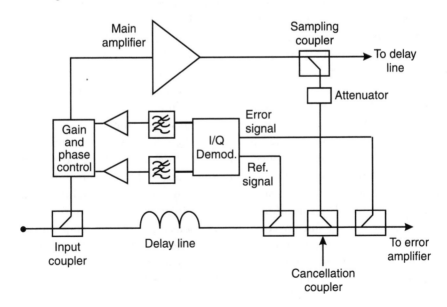

Figure 5.25 Loop 1 analog control.

control networks, and loop balance is obtained when the reference signal is uncorrelated with the error signal (now containing only distortion). One of the advantages of this type of control scheme is that it can compensate for rapid changes, for example, effects due to power ramping; the loop is kept aligned even when the output power is changed. For example, WCDMA uses power control on the downlink (BS to mobile) with power control periods of 0.625 ms and a power stepsize of 1 dB.

5.9.4 Loop 1 Digital Control Example

In the first feedforward loop, the error signal at the output of the cancellation coupler consists of intermodulation from the main amplifier and the suppressed carriers. The relative level of carrier power and intermodulation power depends on the intermodulation performance of the main amplifier (typically −30 dBc for bipolar and −40 dBc for MOSFETs) and the level of loop suppression (target typically >30 dB). Unless the loop suppression exceeds the intermodulation performance of the main amplifier, residual carrier power is the dominant contribution to the error signal. Thus, if the level of residual power could be detected, then it could be used as a control parameter for the first loop; that is, the amplitude and phase are varied until the error power becomes a minimum.

Figure 5.26 shows a possible implementation of this type of control scheme. The error amplifier input signal is first sampled using a directional coupler and, after suitable amplification, a detector is used to generate a dc output signal that is proportional to the level of RF input power. The output of the analog-to-digital converter (ADC) is used together with some form of algorithm to control the amplitude and phase (e.g., an IQ modulator or PIN diodes) until the detected power becomes a minimum.

In its simplest form, the search algorithm simply compares the current and previous values of detected power and then changes the amplitude or phase to see if the detected power increases or decreases. If the power decreases, then the direction of change in amplitude and phase is maintained; alternatively, if the power increases, the direction of change in amplitude and phase is reversed. The algorithm then continues ad infinitum, ensuring that loop balance is maintained over changes in main amplifier gain, temperature, time, and supply voltage. In order to increase the speed with which the algorithm converges to the gain and phase settings for minimum power, more complex search algorithms can be used, for example, Gradient Search methods or techniques using variable stepsize.

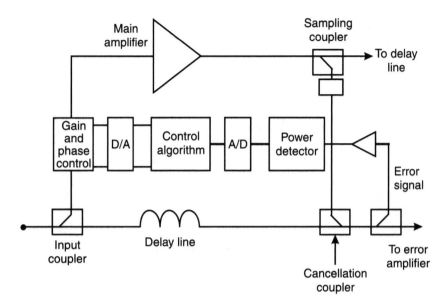

Figure 5.26 Loop 1 digital control.

There are, however, a number of practical issues which need to be considered when using such control schemes. For example, the dynamic range of the power detector, the resolution of the digital/analog (D/A) converters, the speed of convergence, and the shape of the error surface are important factors in determining the overall performance.

5.9.5 Loop 2 Control and Pilot Signals

The situation in the second feedforward loop, which cancels distortion, is different compared to the first feedforward loop, which cancels carriers. In the second loop, the information on loop balance is contained in the distortion rather than the carriers and the relative levels of carrier power and distortion power are typically very large. For example, even when the output distortion is relatively high (e.g., −30 dBc), the carrier power is approximately 1000 times higher than the distortion power. If the distortion level is −60 dBc or even −70 dBc, as required in certain applications, the ratio of "unwanted" carrier power to "wanted" distortion power increases dramatically and it becomes progressively more difficult to detect the distortion. Relatively simple power-detection schemes, such as those explained in the previous section, are thus not practical for distortion-cancellation loops.

One solution to the problem of detecting low levels of output distortion in the presence of high-power carriers is to use a special "pilot" signal, which is suppressed in the same way as distortion in the second feedforward loop. The basic idea is that if the pilot signal is suppressed, then the distortion from the main amplifier is also suppressed.

Figure 5.27 shows one possible implementation; the pilot signal is injected in the first loop and then detected at the feedforward output using the original pilot as a reference. The information is then fed back through loop filters to control the gain and phase such that the output distortion is minimized. Although not shown in Figure 5.27, a second carrier-cancellation loop (sometimes referred to as Loop 3) may also be included to suppress the carriers in the feedback signal.

The exact form of the pilot signal depends upon the particular implementation; for example, the pilot signal could be a narrowband CW signal at

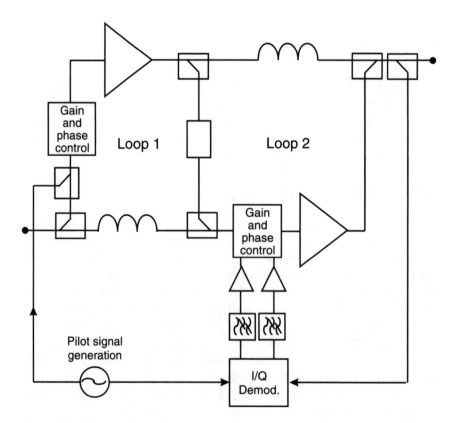

Figure 5.27 Pilot signal control.

a "reserved" frequency or a wideband signal such as a pseudorandom noise sequence. As with Loop 1 control, the goal in Loop 2 is to ensure that the loop remains balanced under all conditions. In some cases however, such as on the peaks of the input signal, loop balance is lost and the intermodulation performance becomes worse.

5.10   Peak Intermodulation

As previously shown (Chapter 3), the intermodulation performance of the main amplifier (e.g., bipolar or MOSFET) is a function of the power level. That is, the intermodulation performance at average power (the average intermodulation) is different compared to the intermodulation performance at peak power (the peak intermodulation). The peak intermodulation of the main amplifier is typically 10 dB higher than the average intermodulation. For a nonconstant envelope signal, such as a multicarrier signal with a Rayleigh distribution, there is thus a corresponding spread in the level of intermodulation.

In a feedforward system, the error amplifier, which is normally designed for average intermodulation levels, saturates on the peaks of the input signal and generates some intermodulation of its own. The second loop also becomes unbalanced during the peaks, thus reducing the level of cancellation for the main amplifier distortion and further degrades the feedforward output linearity. The resultant peak intermodulation at the feedforward output can be as much as 20 dB higher than the average intermodulation level.

Note that the peak intermodulation of a feedforward amplifier can be measured in different ways. For example, statistical analysis can be used (i.e., a certain percentage of the time the peak intermodulation exceeds the average intermodulation by a certain amount) or simple peak hold measurements can be made (the absolute maximum peak level). The peak intermodulation level can be improved by, for example,

- Reducing the peak intermodulation level in the main amplifier;
- Increasing the power-handling capability of the error amplifier.

Alternatively, both the average and peak intermodulation of a feedforward amplifier can be improved significantly, at the expense of extra complexity, by adding additional feedforward loops—that is, double-loop feedforward (Chapter 4).

Appendix

Patented Oct. 9, 1928.   1,686,792

UNITED STATES PATENT OFFICE.

HAROLD S. BLACK, OF NEWARK, NEW JERSEY, ASSIGNOR, BY MESNE ASSIGNMENTS, TO WESTERN ELECTRIC COMPANY, INCORPORATED, A CORPORATION OF NEW YORK.

TRANSLATING SYSTEM.

Application filed February 3, 1925. Serial No. 6,529.

This invention relates to translating systems, and particularly to electric wave amplification.

An object of the invention is to increase the load carrying capacity of such systems.

A related object of the invention is to repeat electrical waves without distortion.

Another object of the invention is to suppress the distortion and modulation produced in an amplifier circuit.

Another object of the invention is to facilitate the operation and maintenance of repeaters in multiplex carrier current signaling systems.

For economic reasons, it is desirable to utilize repeater amplifiers employed in carrier signaling systems to their full load carrying capacity. Whenever the waves impressed upon such an amplifier approach the load limit, however, intermodulation usually results, thereby producing cross-talk which interferes with the efficient reception of signals.

The problem of suppressing cross-talk is complicated by the fact that the cross-talk is a function of the frequency, amplifier output, and the type of modulation. The difficulties heretofore encountered are overcome in the present invention in which both odd and even order products of modulation produced in an amplifier are balanced out for all frequencies, and regardless of the output level of the amplifier.

In the preferred embodiment of the invention, a portion of the current in the input circuit of a repeater amplifier is balanced against a portion of the current in the output circuit. The balance is so adjusted that the frequencies which it is desired to amplify without distortion are balanced out and the cross-talk frequencies alone appear in the output of the balancing circuit. These cross-talk frequencies are amplified and applied to an outgoing line in such phase relation that they just balance out the cross-talk which is transmitted thereto from the repeater output.

The invention will be described as applied to a repeater for multiplex carrier current telephone or telegraph systems, but it is to be understood that it may also be applied to other places in the system than at repeater points, and to various types of systems other than multiplex carrier systems.

In the drawing, Figs. 1 and 2 are circuit diagrams illustrating two different modifications of the invention.

In Fig. 1, a one-way repeater is adapted to couple two line sections W and E. A two-way repeater may be provided by duplicating the circuit to provide a path for operation in the opposite direction.

The signaling waves incoming at the repeater from line section W are amplified in the repeater and the amplified waves are delivered to line section E.

The repeater amplifier may be of any well-known type, such as the general type of balanced amplifier disclosed in the patent to B. W. Kendall, No. 1,544,910, dated July 7, 1925.

As illustrated, the repeater amplifier comprises a pair of three-element electron discharge tubes 1 and 2. The input circuit of the amplifier includes a variable resistance 3, the adjustable taps of which are connected to the grids of the two tubes.

The usual sources of electrical energy 4, 5 and 6 are provided to properly polarize the grid electrodes, to heat the cathodes, and to furnish anode-cathode current, respectively. Each of the anode-cathode direct current circuits of the tubes contains a coil 7 which is designed to offer a high impedance to currents of the frequency or frequencies being repeated, these coils functioning to maintain the sum of the anode-cathode currents of the two tubes constant.

The output circuit of the repeater is connected to the series winding 8 of a differential repeating coil H, commonly known as a hybrid coil. The usual balancing network N is connected to the terminals of the two line windings 9 of the hybrid coil. The signaling waves originating in line section W which are amplified in the repeater, are thus transmitted through the hybrid coil H to the line section E.

Due to the distortion and modulation produced in the repeater, cross-talk frequencies are also impressed upon the line section E, and unless these undesired frequencies are eliminated they may seriously interfere with the efficient reception of the signals at the receiving station.

This invention provides a balancing circuit for eliminating the distortion and modulation frequencies which appear in the output circuit of the repeater. To accomplish

2 1,686,792

this, a portion of the currents in the input circuit of the repeater and a portion of the currents in the output circuit of the repeater are applied to a balancing circuit including a
5 hybrid coil H_1 and its associated balancing network N_1, the hybrid coil H_1 being connected to the input circuit of an amplifier 10, hereinafter called the cross-talk amplifier. The output circuit of the cross-talk amplifier
10 10 is connected to the bridge points of the line windings 9 of the hybrid coil H.

The currents derived from the input of the repeater are applied to the series winding 11 of the hybrid coil H_1, which is
15 coupled to the line windings thereof. The currents derived from the output of the repeater are transmitted through a variable artificial line 12 and applied to the bridge points of the line windings 13 of the hybrid
20 coil H_1. The voltage applied across the series winding 11 of the hybrid coil H_1 is free from distortion and modulation products, while the voltage applied to the line windings 13 of this hybrid coil contains the
25 distortion and modulation components produced in the repeater as well as the undistorted components of the amplified wave.

The amplitude and phase of the voltage which is applied to the line windings 13 of
30 the hybrid coil H_1 are so adjusted by means of the variable artificial line 12 that the frequencies which it is desired to amplify without distortion, which are applied to the series and line windings of the hybrid coil H_1, are
35 exactly equal and balance each other out. Since the frequencies which are free from distortion are balanced out in the hybrid coil H_1, the currents impressed upon the input of the cross-talk amplifier 10 are made
40 up only of the modulation and distortion components produced in the repeater. The gain of the amplifier 10 and the poling of its output are so adjusted that the cross-talk applied to the line windings 9 of the hybrid
45 coil H just neutralizes the cross-talk applied to the series winding 8 from the repeater amplifier.

Fig. 2 illustrates a modification of the invention in which a form of Wheatstone
50 bridge is employed in place of the hybrid coils of Fig. 1 to couple the cross-talk balancing circuit to the input and output circuits of the repeater.

The repeater in this case is similar to that
55 shown in Fig. 1, the additional resistance elements in the input circuit of the push-pull amplifier simply being employed to secure a more perfect balance of the two sides of the circuit. The condensers in the input and
60 output circuits of the repeater may be employed to prevent the flow of any continuous currents through the respective conductors in which they are placed.

A portion of the currents in the input and
65 output circuits of the repeater are applied across the ratio arms of the Wheatstone bridge 14, the arms of which contain equal resistance elements. The output circuit of the repeater and the output circuit of the
70 cross-talk amplifier are coupled to the line section E by means of a similar bridge 15. The variable attenuator 12 may be adjusted until the voltage applied to the Wheatstone bridge 14 from the input side of the repeater is just balanced out by the distortionless
75 components of the voltage applied across the bridge 14 from the output side of the repeater.

As in the case of the system of Fig. 1, the currents impressed upon the input of the
80 cross-talk amplifier 10 are made up only of the modulation and distortion components produced in the repeater. By properly adjusting the gain of the amplifier 10, assuming the output to be properly poled, the
85 cross-talk impressed upon the bridge 15 from this amplifier exactly neutralizes the cross-talk components applied to the bridge from the repeater.

A system constructed in accordance with
90 the invention as described above is very economical to operate and maintain, particularly since the repeater can be operated at a high load level. It will be noted that both odd and even order products of modu-
95 lation are balanced out, and that the balance is independent of the frequency and the output level of the repeater.

The invention set forth herein is, of course, susceptible of various other modifications
100 and adaptations not specifically referred to, but included within the scope of the appended claims.

What is claimed is:

1. The method of suppressing even and
105 odd order distortion components produced in an electric wave translating system and lying within the same frequency spectrum as the undistorted components which comprises isolating a portion of said distortion com-
110 ponents from the system and reimpressing upon the output circuit of said system said isolated distortion currents in opposite phase relation to the even and odd order distortion components therein.
115

2. The method of suppressing distortion components in an electric wave translating system which comprises selecting a portion of the distortion and undistorted components
120 from the output circuit of said system, balancing the undistorted part of said selected components, and reimpressing the distortion components upon said output circuit in phase opposition to the original distortion
125 components therein.

3. The method of suppressing distortion components produced in an electric wave translating system which comprises balancing a portion of the frequencies including
130 the distortion components in said system

against a portion of the frequencies which are free from distortion components to isolate the distortion components, and impressing said distortion components upon the output circuit of said system in opposite phase relation to the original distortion components therein.

4. The method of suppressing distortion components produced in an electric wave amplifier, which comprises balancing a portion of the distortionless current components in the input of said amplifier against a portion of the currents including the distortion components in the output of said amplifier in such phase and amplitude as to neutralize the distortionless components, and impressing the distortion components so derived upon the output circuit of said system in such phase and amplitude as to neutralize the original distortion components therein.

5. A repeater comprising an amplifying element having input and output circuits coupling incoming and outgoing transmission lines, means to shunt a part of the current in the output circuit of said repeater; means for balancing the undistorted frequencies in said shunt circuit to isolate a portion of the distortion components produced in said repeater, and means to impress said distortion components upon said outgoing line in such phase and amplitude as to neutralize the distortion components transmitted thereto from said amplifying element.

6. A repeater comprising an amplifying element having input and output circuits coupling incoming and outgoing transmission lines, means to shunt a part of the current in the output circuit of said repeater, means for balancing the undistorted frequencies in said shunt circuit to isolate a portion of the distortion components appearing in said output circuit, and means to impress said distortion components upon said outgoing line in such phase and amplitude as to neutralize the distortion components transmitted thereto from said repeater.

7. A repeater for currents of different frequencies comprising an amplifying element having input and output circuits coupling incoming and outgoing transmission lines, means for balancing a portion of the frequencies including the distortion components in said repeater against a portion of the frequencies which are free from distortion to isolate the distortion components, and means to impress said distortion components upon said outgoing line in such phase and amplitude as to neutralize the distortion components transmitted thereto from said repeater.

8. A repeater comprising an amplifying element having input and output circuits coupling incoming and outgoing transmission lines, means for balancing a portion

of the distortionless current components in said incoming line against a portion of the currents including the distortion components in the output circuit of said repeater in such phase and amplitude as to neutralize the said distortionless components, and means to impress the distortion components so derived upon said outgoing line in such phase and amplitude as to neutralize the distortion components transmitted thereto from said repeater.

9. A repeater comprising an amplifying element having input and output circuits coupling incoming and outgoing transmission lines over which signaling waves are transmitted, a balancing circuit connected to said incoming line and to the output circuit of said repeater, means in said balancing circuit to adjust the phase and amplitude of the voltage derived from the output circuit of said repeater until the voltage components which are free from distortion and modulation neutralize the voltage components derived from said incoming line, and means for impressing voltage components in the output of said balancing circuit upon said outgoing line in such phase and amplitude as to neutralize the distortion and modulation components transmitted thereto from said repeater.

10. A repeater comprising an amplifying element having input and output circuits, an incoming transmission line associated with said input circuit, a hybrid coil having line windings associated with an outgoing transmission line and a series winding connected to said repeater output circuit, a balancing circuit including a second hybrid coil having line windings and a series winding, means for applying a portion of the distortionless current components in said incoming line to the series winding of said second hybrid coil, means for applying a portion of the currents including the cross-talk components in the output circuit of said repeater to the line windings of said second hybrid coil, means for adjusting the phase and amplitude of the current components applied to the line windings of said second hybrid coil until the voltage components which are free from distortion and modulation are neutralized therein, and means for impressing the voltage components containing distortion and modulation which are isolated in said second hybrid coil to the line windings of said first hybrid coil in such phase and amplitude as to neutralize the distortion and modulation components transmitted thereto from said repeater.

11. A repeater in a multiplex system for currents of different frequencies comprising an amplifier having input and output circuits coupling incoming and outgoing transmission lines, means for shunting a portion of the current in said output circuit, means

4   1,686,792

for balancing the unmodulated current in said shunt circuit to isolate a portion of the modulation components produced in the output circuit, and means for impressing said
5 isolated modulation components upon the outgoing line to neutralize the modulation components transmitted thereto from the amplifier.

12. A repeater in a multiplex system for
10 currents of different frequencies comprising an amplifier having input and output circuits coupling incoming and outgoing transmission lines, means for shunting a portion of the current in said output circuit, means
15 for balancing the unmodulated current in said shunt circuit to isolate a portion of the even and odd order modulation components produced in the output circuit, and means for impressing said isolated even and odd
20 order modulation components upon the outgoing line to neutralize the even and odd order modulation components transmitted thereto from the amplifier.

13. A repeater in a multiplex system for currents of different frequencies comprising 25 an amplifier having input and output circuits coupling incoming and outgoing transmission lines, means for isolating from the unmodulated components a portion of the modulated components produced in the out- 30 put circuit, part of said isolated modulated component lying within the same frequency spectrum as said unmodulated component, and means for impressing said isolated modulated components upon the outgoing line to 35 neutralize the modulation components transmitted thereto from the amplifier.

In witness whereof, I hereunto subscribe my name this 28th day of January A. D., 1925.

HAROLD S. BLACK.

Oct. 9, 1928.   1,686,792

H. S. BLACK

TRANSLATING SYSTEM

Filed Feb. 3, 1925

Fig. 1

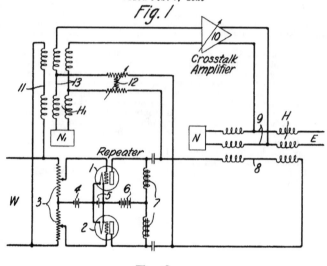

Fig. 2

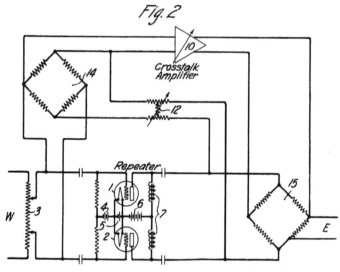

Inventor:
Harold S. Black.
by ℰ.ω.αδ——— Att'y

Bibliography

Bateman, A., "The Combined Analogue Locked Loop Universal Modulator (CALLUM)," *42nd IEEE VTC Conf. Proc.*, Vol. VTC-92, 1992, pp. 759–763.

Bennett, T. J., and R. F. Clements, "Feedforward—an Alternative Approach to Amplifier Linearisation," *The Radio and Electronic Engineer*, Vol. 44, No. 5, May 1974, pp. 257–262.

Black, H. S., 1928, US Patent No. 1,686,792 "Translating System," Filed Feb. 1925.

Boolarian, M., and J. P. McGeehan, "Twin-Loop Cartesian Transmitter," *Electronic Letters*, Vol. 32, No, 11, May 1996, pp. 97–972.

Bowick, C., *RF Circuit Design*, SAMS Books, a division of Prentice Hall Computer Publishing, Indiana, 1996.

Briffa, M. A., and M. Faulkner, "Stability Analysis of Cartesian Feedback Linearisation for Amplifiers with Weak Nonlinearities," *IEE Proc.-Commun.*, Vol. 143, No. 4, Aug. 1996, pp. 212–218.

Cheng, D. K., *Fundamentals of Engineering Electromagnetics*, Reading, MA: Addison-Wesley, 1993.

Dahlman, E., et al., "UMTS/IMT-2000 Based on Wideband CDMA," *IEEE Communication Magazine*, Sept. 1998, pp. 70–80.

Hetzel, S. A., A. Bateman, and J. P. McGeehan, "A LINC Transmitter," *41st IEEE VTC Conf. Proc.*, 1991, pp. 133–137.

Horowitz, P., and W. Hill, *The Art of Electronics,* 2nd Edition, London: Cambridge Univ. Press, 1995.

IEE, "Linear RF Amplifiers and Transmitters," *IEE Colloq.,* London, U.K, April 1994.

Kahn, L., "Single-Sideband Transmission by Envelope Elimination and Restoration," *Proc. I.R.E.,* July 1952, pp. 803–807.

Kahn, L., "Comparison of Linear Single-Sideband Transmitters with Envelope Elimination and Restoration Single-Sideband Transmitters," *Proc. I.R.E.,* Dec. 1956, pp. 1706–1712.

Kennington, P. B., and D. W. Bennett, "Linear Distortion Correction Using a Feedforward System," *IEEE Trans. Vehicular Tech.,* Vol. 45, No. 1, Feb. 1996, pp. 74–81.

Kreyszig, E., *Advanced Engineering Mathematics,* 7th Edition, New York: Wiley, 1993.

Lilliesköld, G., "Olinjära teknik," *Elektronik i Norden,* Vol. 14, 1998, pp. 42–52.

Parsons, K. J., and P. B. Kennington, "Effect of Delay Mismatch on a Feedforward Amplifier," *IEE Proc. Circuits, Devices Syst.,* Vol. 141, No. 2, April 1994, pp. 140–144.

Powell, J., *Linear Amplification for RF Transmitters, Part I and Part II,* Telia SA (France), 1996.

Powell J., et al., US Patent No. 5,323,119 "Amplifier having feedforward correction," 1994.

Sedra, A. S., and K. C. Smith, *Microelectronic Circuits,* 2nd Edition, New York: Holt, Rinehart and Winston, 1987.

Seidel, H., "A Microwave Feedforward Experiment," *Bell Syst. Tech. J.,* Nov. 1971, pp. 2879–2916.

Taub, H., and D. L. Schilling, *Principles of Communication Systems,* 2nd Edition, New York: McGraw-Hill, 1986.

Vizmuller, P., *RF Design Guide,* Norwood, MA: Artech House, 1995.

Wilkinson, R. J., and P. B. Kennington, "Specification of Error Amplifiers for Use in Feedforward Transmitters," *Proc. IEE Proc.-G,* Vol. 139, No. 4, Aug. 1992, pp. 477–480.

About the Author

Nick Pothecary has worked on the design, development, and production of linear power amplifiers at Nokia, Ericsson, and Radio Design (Sweden). He holds a Ph.D. in radio communications from the University of Bristol (UK) and has also worked in the areas of antenna design, radar, and electromagnetic field theory. He has taught many successful courses on linear power amplifiers and electromagnetic theory. His email address is nick.pothecary@telia.com

Index

Microstrip Lines and Slotlines, Second Edition, K.C. Gupta *et al.*

Microwave and Millimeter-Wave Diode Frequency Multipliers, Marek T. Faber *et al.*

Microwave Engineers' Handbook, Two Volumes, Theodore Saad, editor

Microwave Filters, Impedance-Matching Networks, and Coupling Structures, George L. Matthaei, Leo Young, E.M.T. Jones

Microwave Mixers, Second Edition, Stephen Maas

Microwaves and Wireless Simplified, Thomas S. Laverghetta

MULTLIN for Windows: Circuit-Analysis Models for Multiconductor Transmssion Lines, Software and User's Manual, Antonije R. Djordjevic *et al.*

The RF and Microwave Circuit Design Handbook, Stephen A. Maas

RF Design Guide: Systems, Circuits, and Equations, Peter Vizmuller

RF and Microwave Coupled-Line Circuits, Rajesh Mongia, Inder Bahl, and Prakash Bhartia

RF Power Amplifiers for Wireless Communications, Steve C. Cripps

RF Systems, Components, and Circuits Handbook, Ferril Losee

SPURPLOT: Mixer Spurious-Response Analysis with Tunable Filtering, Software and User's Manual, Version 2.0, Robert Kyle

TRANSLIN: Transmission Line Analysis and Design, Software and User's Manual, Paolo Delmastro

Transmission Line Design Handbook, Brian C. Waddell

TRAVIS Pro: Transmission Line Visualization Software and User's Manual, Professional Version, Robert G. Kaires and Barton T. Hickman

For further information on these and other Artech House titles, including previously considered out-of-print books now available through our In-Print-Forever® (IPF®) program, contact:

Artech House
685 Canton Street
Norwood, MA 02062
Phone: 781-769-9750
Fax: 781-769-6334
e-mail: artech@artechhouse.com

Artech House
46 Gillingham Street
London SW1V 1AH UK
Phone: +44 (0)171-973-8077
Fax: +44 (0)171-630-0166
e-mail: artech-uk@artechhouse.com

Find us on the World Wide Web at: www.artechhouse.com